# The Psychology of Learning Mathematics

Paul Ernest

# The Psychology of Learning Mathematics

The cognitive, affective and contextual domains of mathematics education

LAP LAMBERT Academic Publishing

**Impressum/Imprint (nur für Deutschland/ only for Germany)**
Bibliografische Information der Deutschen Nationalbibliothek: Die Deutsche Nationalbibliothek verzeichnet diese Publikation in der Deutschen Nationalbibliografie; detaillierte bibliografische Daten sind im Internet über http://dnb.d-nb.de abrufbar.

Alle in diesem Buch genannten Marken und Produktnamen unterliegen warenzeichen-, marken- oder patentrechtlichem Schutz bzw. sind Warenzeichen oder eingetragene Warenzeichen der jeweiligen Inhaber. Die Wiedergabe von Marken, Produktnamen, Gebrauchsnamen, Handelsnamen, Warenbezeichnungen u.s.w. in diesem Werk berechtigt auch ohne besondere Kennzeichnung nicht zu der Annahme, dass solche Namen im Sinne der Warenzeichen- und Markenschutzgesetzgebung als frei zu betrachten wären und daher von jedermann benutzt werden dürften.

Coverbild: www.ingimage.com

Verlag: LAP LAMBERT Academic Publishing GmbH & Co. KG
Dudweiler Landstr. 99, 66123 Saarbrücken, Deutschland
Telefon +49 681 3720-310, Telefax +49 681 3720-3109
Email: info@lap-publishing.com

Herstellung in Deutschland:
Schaltungsdienst Lange o.H.G., Berlin
Books on Demand GmbH, Norderstedt
Reha GmbH, Saarbrücken
Amazon Distribution GmbH, Leipzig
**ISBN: 978-3-8443-1306-2**

**Imprint (only for USA, GB)**
Bibliographic information published by the Deutsche Nationalbibliothek: The Deutsche Nationalbibliothek lists this publication in the Deutsche Nationalbibliografie; detailed bibliographic data are available in the Internet at http://dnb.d-nb.de.

Any brand names and product names mentioned in this book are subject to trademark, brand or patent protection and are trademarks or registered trademarks of their respective holders. The use of brand names, product names, common names, trade names, product descriptions etc. even without a particular marking in this works is in no way to be construed to mean that such names may be regarded as unrestricted in respect of trademark and brand protection legislation and could thus be used by anyone.

Cover image: www.ingimage.com

Publisher: LAP LAMBERT Academic Publishing GmbH & Co. KG
Dudweiler Landstr. 99, 66123 Saarbrücken, Germany
Phone +49 681 3720-310, Fax +49 681 3720-3109
Email: info@lap-publishing.com

Printed in the U.S.A.
Printed in the U.K. by (see last page)
ISBN: 978-3-8443-1306-2

Copyright © 2011 by the author and LAP LAMBERT Academic Publishing GmbH & Co. KG and licensors
All rights reserved. Saarbrücken 2011

# TABLE OF CONTENTS

| | |
|---|---|
| **INTRODUCTION** | 3 |
| **CHAPTER 1: TYPES OF RESEARCH IN MATHEMATICS EDUCATION** | 5 |
| **CHAPTER 2: THE AIMS AND OBJECTIVES OF LEARNING MATHEMATICS** | 9 |
| **CHAPTER 3: LEARNING FACTS AND SKILLS IN MATHEMATICS** | 21 |
| **CHAPTER 4: CHILDREN'S ERRORS AND METHODS IN MATHEMATICS** | 32 |
| **CHAPTER 5: CHILDREN'S LEARNING OF LEARNING OF CONCEPTS AND STRUCTURES** | 47 |
| **CHAPTER 6: THEORETICAL & PRACTICAL ASPECTS OF CONSTRUCTIVISM** | 63 |
| **CHAPTER 7: MATHEMATICAL PROCESSES AND STRATEGIES** | 72 |
| **CHAPTER 8: THE AFFECTIVE DOMAIN** | 100 |
| **CHAPTER 9: THE CONTEXT OF LEARNING MATHEMATICS** | 120 |
| **CHAPTER 10: MATHEMATICAL KNOWLEDGE AND CONTEXT** | 133 |
| **REFERENCES** | 148 |

# INTRODUCTION

This book addresses the research and theory of the psychology of learning and teaching mathematics.

The aims of this book are to:
- Introduce readers including teachers and educational professionals to current thinking in the psychology of mathematics education.
- Bring readers up to date on research, developments and theories in the psychology of learning mathematics.
- Enable readers to look more closely at student understanding of mathematics, performance in assessment of the knowledge, and the implications for teaching, across the mathematics curriculum.

## OVERVIEW

This book provides an introduction to and an overview of current thinking, research and literature on the psychology of learning mathematics. The main organising framework employed is the classification of the central learning outcomes of school mathematics into facts, skills, conceptual structures, general strategies and appreciation (based on Bell *et al.* 1983) Chapter 2 treats the aims and objectives of learning mathematics, including Bell and colleagues' framework. The learning of facts and skills in mathematics is treated in Chapters 3 and 4, with the latter focussing on children's errors and methods in mathematics. Understanding, conceptual structures and the construction of meaning is the subject of chapters 5 and 6, with the former exploring children's learning of concepts and conceptual structures in mathematics, and the latter exploring theoretical and practical aspects of constructivist learning theory. Chapter 7 treats mathematical processes and strategies, the thought processes involved in mathematical problem solving. Finally, in fleshing out the organising framework, chapter 8 treats attitudes to and the appreciation of mathematics. However, this is not the end of the work because a further theme has emerged as important in psychology of mathematics education in recent years. This is the context of learning mathematics

Two other themes form the basis of the first and last chapters. First of all, the text begins in Chapter 1 with a brief account of research methodology, and in particular, the two main types of research approaches used in mathematics education, and indeed throughout educational and social science research. This material is included to raise awareness of the multiple research methods and methodologies in use in research in the psychology of learning mathematics, and to focus readers' awareness on the means of gathering and validating research evidence which is a significant (if implicit) theme in all chapters. It is treated thoroughly in a further forthcoming book on research methods and methodologies in mathematics education (Ernest forthcoming d).

The second additional theme is treated in the last two chapters of this book, namely the context of teaching and learning mathematics. This raises the issues of the impact of the social context on the teaching and learning of mathematics, and the cognitive and psychological significance of both the external representation of mathematical ideas and tasks and also their internal (mental) representations, as well as theories of the links between the two. Chapter 9 treats the context of learning mathematics including both task context and social context. Chapter 10 goes on to address the more theoretical issues of mathematical knowledge and context, including the important issue of the transfer of knowledge between contexts.

The text does not claim to provide an exhaustive coverage of the psychology of learning mathematics. It constitutes an introduction to the central topics in the area. There are further psychological themes and issues that are not treated here, or only incidentally, and which the interested reader will need to explore on her/his own. Some of the topics missing or treated very briefly are the following.

First of all, there is no detailed study of the psychology of the learning of specific mathematical topics, except by way of illustration. Research like this has been an important growth area in the past two decades or so. Surveys can be found in APU (1986), Bell *et al.* (1983), Biehler *et al.* (1994), Bishop (1996), Dessart and Suydam (1983), Dickson *et al.* (1984), Ernest (1996), Grouws (1992), Hart (1981), Jensen (1993), Owens (1993), and Wilson (1993). In addition there are monographs specifically treating a wide range of topics such as the psychology of learning number, algebra, geometry, advanced mathematical thinking, etc. (e.g., Wagner and Kieran 1989).

Second, a further missing cluster of areas includes different teaching and learning styles, brain localisation and specialisation, and the different modes of thought associated with the left and right cerebral hemispheres. To some extent, because of the claims associating different thinking styles and brain hemisphere functions with gender differences, this topic is treated (albeit briefly) in a further forthcoming book on Gender, equal opportunities and the nature of mathematics (Ernest forthcoming c).

A third missing area is that of language and mathematics, especially the problem areas of dyslexia, dyscalculia, specific learning difficulties and disabilities in learning mathematics, and more generally, special educational needs in mathematics. This is treated in a further forthcoming book on Mathematics and Special Educational Needs (Ernest forthcoming a).

A fourth cluster of areas which used to be central to the psychology of learning mathematics but has become more peripheral is that of Piagetian theory, and critiques of learning hierarchies and hierarchical theories of mathematical ability. This is barely treated at all here. There is some treatment of this area in sections 2, 3 and 4. See also, e.g., Brown and Desforges (1979), Ernest (1991a), Ernest (1996) and Walkerdine (1984) for a discussion of some of the issues.

A fifth set of ideas relevant to the psychology of learning mathematics is the structuring of the mathematics curriculum and the use of psychology in the development of mathematical assessment. Although this is more of an application of the psychology of learning mathematics than a subfield of it, it is an important area of research, policy and practice. This is treated in a further forthcoming companion book on the Mathematics curriculum and assessment (Ernest forthcoming b)

Lastly, there is the psychological impact of computers, calculators, and information and communication technologies in general, on the teaching and learning of mathematics. Although there is some interesting work on psychological dimensions of this area, it has not been treated here. The interested reader can look at, e.g., Biehler *et al.* (1994), Ernest (1989), Ernest (1996), Hoyles and Noss (1992), Hoyles and Sutherland (1989), Jensen (1993), Lindquist (1988), Noss (1988), Papert (1980) and Wilson (1993).

Thus although this book focuses on the central areas of research and knowledge in the psychology of learning mathematics it does not claim to provide an exhaustive map of the field. Some indication of the breadth of the field is provided by headings used to divide up references in Ernest (1996). The psychology of mathematics education is undoubtedly the largest area of research within mathematics education, and there are many hundreds of papers and books published annually in this specialist area including international conference proceedings and journals in the mathematics education and the psychological research communities.[1]

---

[1] The *International Group for the Psychology of Mathematics Education* (PME) holds an annual conference and publishes its proceedings which includes several hundred research papers in this area of research.

# CHAPTER 1

## TYPES OF RESEARCH IN MATHEMATICS EDUCATION

Recently, there has been a shift in the accepted styles of research on the learning of mathematics from the scientific approach, which focuses on measurable outcomes and generalisable results, to the interpretative approach, which explores a small number of cases in depth, to illuminate the mechanisms at work in those cases. As in psychology, once the scientific research approach was regarded as the only legitimate way to explore the learning of maths, but much research since the 1980s has employed the interpretative approach.

Since most mathematics teachers have a scientific background, they often think that the scientific approach is the only way to do 'real' research. To be rigorous, a course assignment should use proper experimental designs with random distribution of subjects to experimental and control groups, whose learning is measured through the collection of quantitative data using predetermined questionnaires, tests, etc. The goal of such research is objective knowledge in terms of testable hypotheses about behaviour which are expressed mathematically. What is often overlooked is the weakness of this approach. The uniqueness of individuals, classes and schools, and the limited samples teachers have the time to investigate rarely allow full generalisation. There are often unquestioned assumptions in the enquiry about how to measure learning or understanding. In addition, statistical tests of significance often require assumptions about the regularity of the distribution of underlying variables. Even when valid, statistical data does not tell about the causal connections between variables, nor what sense learners make of the teaching and learning situations.

For reasons such as these, as well as the influence of research in sociology and the humanities, much research is now adopting an interpretative research approach, sometimes also called the qualitative research style. This approach has been finding favour with a growing number of researchers interested in how students learn mathematics. It is primarily about generating case studies of learners and learning, finding out student's personal understandings about what is going on in the learning process. Table 1.1 below summarises the two approaches.

Both styles of research are legitimate, and have a place (and strengths and weaknesses). Both types of data will be used on this course. We learn about thinking by looking at cases of children making errors. We also learn by looking at test results. Together the two kinds of data combine to give a better picture.

*Table 1.1: Contrasting Scientific and Interpretative Research approaches*

| Style | Scientific Research | Interpretative Research |
|---|---|---|
| **Origins** | Scientific research in physics, biology, agriculture, psychology, etc. | Research in sociology and the humanities |
| **Goal** | Objective knowledge of scientific truths, Replicable laws of behaviour which can be expressed mathematically | Illuminating personal truths, Telling case studies of learners and learning, Subjective understanding, Textual descriptions of learning |
| **Methods** | Collection of primarily quantitative data through predetermined questionnaires, tests, etc. Uses experimental designs (experimental vs. control groups, etc.) | Collection of primarily qualitative data through interviews, detailed case studies of small samples of learners and classrooms (sometimes just one) |
| **Strengths** | Objective scientific knowledge which is generalisable and replicable; Valid and reliable data. Comparisons between samples possible. Relates to the assessment of learning | Uses methods responsive to individuals and context. Illuminating studies of unique individuals and situations. 'inside story'; rich descriptions of mechanisms linking different observed factors |
| **Weaknesses** | Samples rarely allow full generalisation. Unquestioned assumptions common. Causal links unspecified. | Risk of subjectivity of inquiry and results, anecdotal outcomes. Non-generalisability, lack of predictions. |

In conducting mini investigations for course assignments, teachers often find it easier and more fruitful to look at a few learners in depth (interpretative type research) and are rarely able to do proper teaching experiments with control groups etc. (scientific type research). However both approaches can and are used successfully. Indeed, both types of methods can be successfully combined in an appropriate investigation.

From the perspective of this module, it is important for course members to note that their assignments can validly investigate a small number of learners' responses to routine or problem solving tasks, using interviews or diagnostic assessments in what are typically interpretative research approaches. Details of the possible range of coursework assignments are given in a separate handout.

Chapter 4 below, which is about errors and children's methods compares and combines data derived from both an interpretative study of one boy's understanding of fractions with a large scale scientific survey of children's errors in adding fractions. This illustrates the importance, significance, and validity of the insights derived from both types of research.

This brief introduction is merely intended to raise awareness of some of the issues involved. More details about the different research approaches are given in the module on Research Methodology in Mathematics Education. All assignments should be written with some self-awareness about research methodology and for course members wishing to know more at this stage the following references are recommended.

## Recommended Readings

Bassey, M. (1990-91) On the Nature of Research in Education (Parts 1-3), *Research Intelligence* Nos. **36** (Summer 1990), 35-38; **37** (Autumn 1990), 39-44; **38** (Winter 1991) 16-18. This set of three articles which provides a valuable if brief introductory overview of many of the issues that arise in research methodology course.

Bell, J. (1992) *Doing Your Research Project*, Milton Keynes, Open University Press. A clear and straightforward introduction to different approaches to educational research.

Cohen, L. and Manion, L. (1995) *Research Methods in Education*, London: Methuen (4th edition). A standard text on research methods in education which students have often found useful.

Ernest, P. (1994) *An Introduction to Educational Research Methodology and Paradigms,* Exeter: University of Exeter School of Education (Graduate School). An introduction to theoretical and practical issues in research methodology.

Ernest, P. (Forthcoming d) *Research Methodology in Mathematics Education*, Saarbrücken, Germany: Lambert Academic Publishing. This treats many of the theoretical ideas underpinning the use of research methods and methodologies in mathematics and science education.

McNiff, J. (1988) *Action Research: Principles and Practice*, London: Macmillan. An accessible account of the approach with examples of teacher's case study research.

Silverman, D. (1985) *Qualitative Methodology and Sociology*, Aldershot: Gower Publishing Company. This provides a well grounded introduction to ethnographic and qualitative research methods, located in the interpretative research paradigm. It discusses sociological and social science research, and is not specifically directed at education, but a number of educational researchers have found it to be of value.

# CHAPTER 2

## THE AIMS AND OBJECTIVES OF LEARNING MATHEMATICS

### The Aims of Teaching and Learning Mathematics.

The teaching of mathematics is an intentional activity. It is the result of extensive professional experience and careful planning, at the teacher's, the departmental, school, district, curriculum developer's and the national levels. The underpinning intentions, stated in terms of their intended purposes and outcomes, constitute the aims of education. A number of different terms are used to refer to these outcomes including aims, goals, targets and objectives. Since Taba (1962) a distinction is commonly drawn in education between short term educational goals (objectives) and 'broad aims', which are longer term and less specific goals (aims). I shall follow this usage, employing the terms 'aims' and 'goals' for broader, longer term educational ends and 'objectives' and 'targets' for narrower, shorter term ends.

Many statements of the aims of the teaching and learning of mathematics have been published. For example, in 1985, in the wake of the Cockcroft Report (1982), HM Inspectors of Schools specified the following set of aims.

> The aims of mathematics teaching
> 1.1 There are important aims which should be an essential part of any general statement of intent for the teaching of mathematics. Those stated in this chapter are considered to be indispensable but it is recognised that there may be others which teachers will wish to add. These aims are intended for all pupils although the way they are implemented will vary according to their ages and abilities.
> [Pupils should have some mastery and appreciation of]
> 1.2 Mathematics as an essential element of communication
> 1.3 Mathematics as a powerful tool  [they should develop]
> 1.4 Appreciation of relationships within mathematics
> 1.5 Awareness of the fascination of mathematics
> 1.6 Imagination, initiative and flexibility of mind in mathematics
> [they should acquire the personal qualities of]
> 1.7 Working in a systematic way
> 1.8 Working independently
> 1.9 Working co-operatively  [and two further desired outcomes are]
> 1.10    In-depth study of mathematics
> 1.11    Pupils' confidence in their mathematical abilities"
>                (Her Majesty's Inspectorate 1985: 2-6)

Elsewhere I have argued that different social groups have different aims for the teaching of mathematics. In Ernest (1991a) I distinguish the aims of five groups in modern Britain, each of which emphasises one of the following set of aims:

*Table 2.1: Contrasting aims of five groups in modern Britain*

- Learning to solve practical problems with mathematics (work centred)
- Understanding and appreciating maths (mathematics centred)
- Confidence, creativity and self expression through maths (child-centred)
- Empowerment of learner as highly numerate critical citizen in society (socially aware)
- Acquiring basic mathematical skills (basic skills centred)

Aims are closely tied in with ideologies and philosophies of education, and this link is explored more fully in the advanced module on the Mathematics Curriculum. For a review of the aims and goals of mathematics education see Niss (1996). Here I want to focus on the more specific psychological learning outcomes; the objectives of learning mathematics.

## The Objectives of Learning Mathematics

In addition to aims there are the specific objectives of the teaching and learning of mathematics. A tradition of specifying the learning objectives goes back a long way. In the 1950s Benjamin Bloom (1956) and colleagues developed a hierarchical theory of learning objectives. In accordance with the educational and cognitive psychology of the day, they made the important distinction between three different domains of psychological functioning::

1. The *cognitive* domain, concerned with knowledge, understanding, reasoning and intellectual functions;

2. the *affective* domain, concerning attitudes, feelings, and values; and

3. the *psycho-motor* domain, concerning physical skills and dexterity and perceptual skills.

The objectives of learning mathematics include each of these domains, because attitudes to mathematics (domain 2) are important, as is the ability, for example to use a pair of compasses, a protractor, or a ruler (domain 3).

Bloom's Taxonomy of the cognitive domain is made up of 6 levels of skills at different levels of complexity and intellectual demand. Thus objectives can be located at these different levels

*Table 2.2: Bloom's Taxonomy of the Cognitive Domain*

| LOWER LEVELS | 1 | **KNOWLEDGE:** Recall of knowledge items |
|---|---|---|
| | 2 | **COMPREHENSION:** translation, interpretation, extrapolation |
| | 3 | **APPLICATION** use of abstractions in concrete situations |
| | 4 | **ANALYSIS** of elements, relationships, organisational principles |
| | 5 | **SYNTHESIS:** production of unique communications, plans, sets of abstract relations |
| HIGHER LEVELS | 6 | **EVALUATION** judgements using internal evidence, external criteria |

Level 1 (recall of knowledge items) is the lowest level of cognitive operating (e.g. recalling Sin $a$ = O/H)

Levels 2 and 3 are slightly higher levels of intellectual functioning, concerned with understanding a mathematical task or situation and acting on it in some way, such as continuing a sequence (e.g. 7, 10, 13, 16,...) or applying a rule.

Levels 4 and 5 (analysis and synthesis) are higher levels of intellectual functioning, mainly concerned with problem solving (e.g. analysing table of results such as 7, 10, 13, 16,... for pattern $3n+4$).

Level 6 is the highest level of cognitive functioning, concerned with matters of proof and judgement in mathematics (e.g. finding error in $x^2+2x+1 = 5x^2+x-4 \Rightarrow (x+1)^2 = (5x-4)(x+1) \Rightarrow x+1 = 5x-4 \Rightarrow 4x = 5 \Rightarrow x = 5/4$ is a simple example.)

The examples given are rather simple and are only intended to apply to appropriately young secondary school students. The cognitive demand of the items indicated is a function of the student's knowledge, and hence of her age and attainment level.

Freudenthal published a powerful critique of this taxonomy on several grounds. First of all, it does not apply well to mathematics. Second, it is really about assessment systems and not about learning. Thirdly, it assumes knowledge and skill can be analysed into lots of different pieces, when problem solving skill involves integrating many types of knowledge and skill in a single act of understanding. Therefore I warn readers against using this outmoded framework without first acquainting themselves with Freudenthal's (1978) critique (see pages 81-92, etc).

More recently, Alan Bell and colleagues (1983) in a review of the psychological research on the learning of mathematics commissioned for the Cockcroft Report (1982) offered a breakdown of the types of learning outcomes in school mathematics, as is shown in the following table

*Table 2.3: The learning outcomes of school maths*

| OUTCOME | DEFINITION |
|---|---|
| FACTS | Items of information that are essentially arbitrary (notation, conventions, conversion factors, names of concepts) |
| SKILLS | well-established multi-step procedures whether involving symbolic expressions, geometrical figures, etc. |
| CONCEPTS | A concept is, strictly speaking, a set or property, a means of discrimination (e.g. concept of negative number) |
| CONCEPTUAL STRUCTURES | a set of concepts and linking relationships - e.g. concept of place value |
| GENERAL STRATEGIES | Procedures which guide choice of what skills or knowledge to use at each stage in problem solving etc. |
| APPRECIATION | of the nature of mathematics and affective response (attitudes) to it |

These learning outcomes represent the knowledge, skill or dispositions acquired by learners when the outcome is successfully achieved. Together, these learning outcomes represent an analysis of the types of knowledge that it is intended that learners should acquire through mathematics teaching. Although each of the categories is treated in greater depth in the following chapters, here is a brief introduction to the knowledge categories.

**Facts**

Facts are items of information that just have to be learned to be known, such as Notation (e.g. the decimal 'point' in place value notation; '%'); Abbreviations (e.g., cm stands for centimetre); Conventions (e.g., 5x means 5 times x; knowing the order of operations in brackets); Conversion factors (e.g., 1 km = 5/8 mile); Names of concepts (e.g., odd numbers; a triangle with three equal sides is called equilateral); Factual results (e.g. multiplication table facts, Pythagoras' rule).

Facts are the basic 'atoms' of mathematical knowledge. Each is a small and elementary piece of knowledge. Facts must be learned as individual pieces of information, although they may fit into a larger more meaningful system of facts. When they fit in this way they are much easier and better remembered, but then they become part of a conceptual structure. For example, 9×6=54 is a fact. But when a child also knows that 9×6=6×9, and that 9×7 has one more ten and one less unit, and 9×4 has one less ten and one more unit, and 9×6=(10-1)×6, and so on, this fact is part of that child's conceptual structure.

## Skills
Skills are well-defined multi-step procedures. They include familiar and often practised skills such as basic number operations. They can involve doing things to numbers (e.g., column addition), or to algebraic symbols (e.g., solving linear equations), or geometrical figures (e.g., drawing a circle of given radius with compasses), etc.

Skills are most often learned by examples: first seeing worked examples, and then 'doing' some. That is, repeated practice of the skill, usually on examples of graduated difficulty.

Seeing how learners actually perform skills is a valuable lesson. For as well as learning skills, children make errors, often on the way to learning the skills. Many of these errors are part of a repeated pattern. They often seem to come from children learning some of the parts of the skill but missing out a part, or putting them together incorrectly. Other errors come from misapplying a rule. For example, in adding fractions, many children simply add the top numbers together, and the bottom numbers. Researchers found that about 20% of secondary school children made the following mistake: 1/3 + 1/4 = 2/7 (See Hart 1981). Why should they do this? It seems likely that they are misapplying the easier multiplication rule for fractions, but adding instead of multiplying.

Error patterns in skills suggest that children absorb some of the different components they have been taught, and put them together in their minds in their own individual ways. This leads to the important conclusion that children themselves construct their skills and knowledge, based on their teaching and learning experiences, and is called the constructivist theory of learning. This also explains how some children invent their own correct but unusual skills.

## Concepts and Conceptual structures
A concept, strictly speaking, is a simple set or property. This is a means of choosing among a larger class of objects those which fit under the concepts. For example, the concept red picks out those objects that we see which are red in colour. The concept of negative number picks out those numbers less than zero. The concept square picks out just those plane shapes which have four straight equal side sides and four equal (right) angles. A concept is the idea behind a name. To learn the name is just to learn a fact, but to learn what it means, and how it is defined, is to learn the concept.

A conceptual structure is a set of concepts and linking relationships between them. It is complex and continues to grow as the child adds more concepts and links through learning. For example, 'place value' and 'quadrilateral' are conceptual structures. Place value is the system of numeration we use which sets the value of a digit, e.g., 9, according to its position or placing. So 9 in the units, tens, hundreds and tenths place has the value 9, 90, 900, 0.9, respectively, with zeros and the decimal point showing its position in the decimal columns. Understanding place value means knowing this, and that each column is worth ten times more than its right hand neighbour, and a tenth as much as its left hand neighbour. So multiplication by 10, 100, 1000 means moving the whole number train (all the digits in a number) one, two or three places, respectively, to the left. It also means knowing that there is no end to the supply of places to the left and right, and that numbers of any size can be expressed with ten digits and a dot.

'Quadrilateral' makes up a simpler conceptual structure. But it includes knowing the relationships between polygons, quadrilaterals, trapeziums, rhombuses, parallelograms, rectangles, squares, and kites.

The conclusion that children construct their own knowledge (this theory is termed 'constructivism', and is discussed in a subsequent chapter) applies even more to conceptual structures. Our memory of all that happens to us, both in and out of school, is put together in a unique way in our mind. I have certain pictures I associate with the numbers 1 to 100, but other people will have different pictures, or other feelings or associations. In other words, our conceptual structures for whole number are different. Of course they should share some features, such as the fact that 11 comes before 12.

Some researchers drew a very thorough map of the basic knowledge and skills making up two digit subtraction, with about 50 components, not counting individual number facts such as 5-3=2 (see Denvir and Brown 1986). They tested quite a few primary school children and found that although the map was a useful tool in describing personal knowledge patterns, it didn't help predict what the children would learn next, even given teaching targeted very carefully at specific skills. Many children did not learn what they were taught, but more surprisingly learned what they were not taught! This fits with the constructivist theory that children follow their own unique learning path and construct their own personal conceptual structures.

Most of the mathematical knowledge that children learn in school is organised into conceptual structures, and the facts and skills they learn can also be fitted in or linked with them. The more connections children make between their facts, skills and concepts the easier it is for them to recall the knowledge and to use and apply it.

**General strategies**

Solving problems is one of the most important activities in mathematics. General strategies are methods or procedures that guide the choice of which skills or knowledge to use at each stage in problem solving.

Problems in school mathematics can be familiar or unfamiliar to a learner. When a problem is familiar the learner has done some like it before and should be able to remember how to go about solving it. When a problem has a new twist to it, the learner cannot recall how to go about it. This is when general strategies are useful, for they suggest possible approaches that may (or may not) lead to a solution. Open ended problems or investigations may require the learner to be creative in exploring a new mathematical situation and to look for patterns.

The first area in which most children learn general strategies is in solving number problems. If asked to add 15 and 47 mentally, children learn to look for ways to simplify the problem. Thus they will often try to make ten with part of the units. They might take 3 from the 5 to add to the 7 to make 10 (15+47=12+50=62) or they might take 5 from the 7 to add to the 5 to make 10 (15+47=20+42=62). Some will simply add the tens and units separately (15+47=50+12=62). The general strategy is that of simplifying the problem through decomposing and recombining numbers.

General strategies are usually learned by example, or are invented or extended by the learner. They are recognised as important in the National Curriculum for children of all ages, and the first attainment target Using and Applying Mathematics is mainly concerned with developing and using general strategies. Three types of general strategy are included in National Curriculum mathematics. The first is "making and monitoring decisions to solve problems" concerning the choice of materials, procedures and approaches in problem solving. The second is "developing mathematical language and communication" which concerns the oral communication and written recording and presentation of problem solving and its results. The third is "developing skills of mathematical reasoning" concerning mathematical thinking, and the use of reasoning to arrive at, check and justify mathematical results.

## Attitudes

Attitudes to mathematics are the learner's feelings and responses to it, including like or dislike, enjoyment or lack of it, confidence in doing mathematics, and so on. The importance of attitudes to mathematics is widely accepted, and one of the common aims of teaching mathematics is that after study, all learners should like mathematics and enjoy using it, and should have confidence in their own mathematical abilities. As well as being a good thing in itself, a positive attitude often lead to greater efforts and better attainment in mathematics. However, too many youngsters and adults sadly say that they dislike mathematics and lack confidence in their abilities. Some even feel anxious whenever it comes up.

Attitudes to mathematics cannot be directly taught. They are the indirect outcome of a student's experience of learning mathematics over a number of years. However, sometimes a particular incident can change a student's attitude, such as teacher encouragement and interest in the learner's work (positive effect), or public criticism and humiliation of the learner in mathematics (negative effect). However these effects are unpredictable, and they depend on the learner's own response to the situation.

## Appreciation

The appreciation of mathematics concerns understanding the 'big picture' of mathematics. It involves some awareness of what mathematics is as a whole (the inner aspect), as well as some understanding of the value and role of mathematics in society (the outer aspect). Appreciation is discussed in greater detail in a later chapter.

Table 2.4 is an activity on which to test out your understanding of these categories.

*Table 2.4: Task applying learning outcomes of school maths*

| **Decide which of the following are best described as: facts, skills, conceptual structures, general strategies, attitudes to, or appreciation of mathematics** |
|---|
| 1. Can reflect right angled triangle in y-axis |
| 2. Knows cm stands for centimetre |
| 3. Says maths is the study of pattern |
| 4. Knows 5/8 mile is 1 km |
| 5. Can calculate that +4 --5 =+9 |
| 6. Knows how to find Tan $x$ for $0°<x<180°$ with calculator. |
| 7. Knows how to find $x$ given values for a, b, c and a:x::b:c (a is to $x$ as b is to c) |
| 8. Can find number of ways of getting a total score of 10 on 3 dice |
| 9. Knows $a.bcd \times 100 = abc.d$ |
| 10. Knows how to test the formula $x^2 + 2x - 17$ to see if it always give prime numbers. |
| 11. Comes back at lunch time to work on GCSE maths project |
| 12. Can identify prime numbers |
| 13. Knows $2x^2$ is $2 \times x \times x$ |
| 14. Says maths is the subject in which you always know if you're right or wrong |
| 15. Says maths is a subject of exciting challenge problems |
| 16. Knows 75% = 3/4 |
| 17. Can draw equilateral triangle with compasses. |
| 18. Decides to test multiples of six to see if they are perfect numbers. |
| 19. Knows 0.7 > 0. 25 |
| 20. Will not do matrix multiplication |
| 21. Prefers 100 simple additions to finding numbers expressible with four 4s combined |
| 22. Can estimate angles to $20°$. |
| 23. Can explain when 9.7 + 8.65 might be used |
| 24. Can calculate total area of classroom walls, floor and ceiling. |
| 25. Knows Tan $a$ = O/A |

## Cockcroft 243 and the National Curriculum

The distinction between the different learning outcomes of facts and skills, conceptual structures, general strategies and appreciation in school mathematics was taken up in the Cockcroft Report (1982). These elements, the report stressed, involve distinct aspects of teaching, and require separate attention.

Thus on purely scientific grounds it can be said that it is not sufficient to concentrate on children's acquisition of facts and skills, if numeracy, understanding, and problem solving ability are the desired outcomes of mathematics teaching. This contradicts the more extreme claims of supporters of the Back-to-Basics movement in education.

On the basis of this psychological research the Cockcroft Report made its most famous recommendation.

Mathematics teaching at all levels should include opportunities for
\* exposition by the teacher;

* discussion between teacher and pupils and between pupils themselves;
* appropriate practical work;
* consolidation and practice of fundamental skills and routines;
* problem solving, including the application of mathematics to everyday situations;
* investigational work. (Cockcroft 1982, paragraph 243)

The teaching approaches needed to develop the elements listed above at all levels of schooling include investigational work, problem solving, discussion, practical work, exposition by the teacher, as well as the consolidation and practice of skills and routines. The following figure 2.1 shows how these teaching approaches contribute to the development of children's appreciation of mathematics, strategies for tackling new problems, conceptual structures in mathematics, as well as their knowledge of mathematical facts and skills. Thus a recognition of the variety of different psychological objectives of school mathematics is what lay behind the reports most famous recommendation.

*Figure 2.1: The Relation between Teaching Styles and Learning Outcomes*

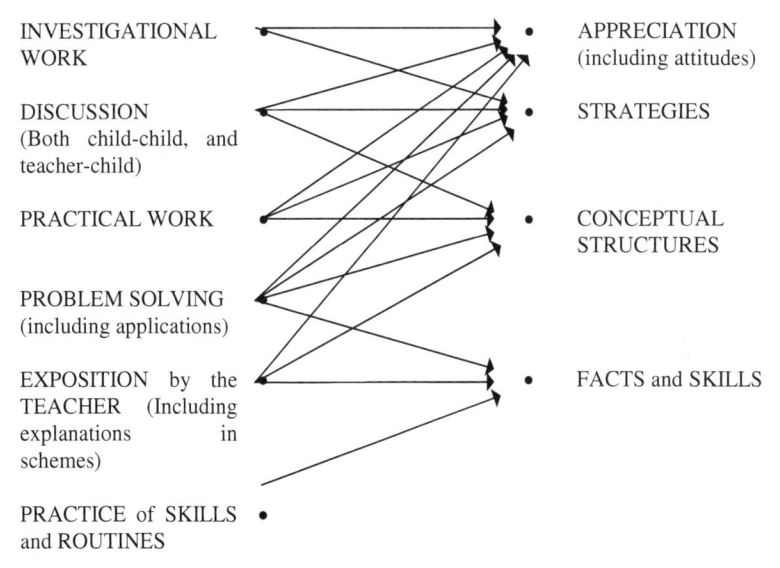

INVESTIGATIONAL WORK → APPRECIATION (including attitudes)

DISCUSSION (Both child-child, and teacher-child) → STRATEGIES

PRACTICAL WORK → CONCEPTUAL STRUCTURES

PROBLEM SOLVING (including applications)

EXPOSITION by the TEACHER (Including explanations in schemes) → FACTS and SKILLS

PRACTICE of SKILLS and ROUTINES

Following the Cockcroft Report a significant official British publication on the teaching of mathematics (HMI 1985) stressed the importance of addressing the full range of mathematics learning objectives, and elaborated them more fully than earlier publications. This is shown in Table 2.5.

*Table 2.5: The Objectives of School Mathematics*

| Facts | Terms (odd, prime) |
|---|---|
| | Notation (place value, %) |
| | Conventions (brackets, mph) |
| | Results (× facts, Pythagoras) |
| **Skills** | Performing basic operations |
| | Sensible use of calculator |
| | Simple practical skills in maths |
| | Ability to communicate maths |
| | Use of micro in maths activities |
| **Conceptual Structures** | Understanding basic concepts |
| | Relationships between concepts |
| | Selecting appropriate data |
| | Using maths in context |
| | Interpreting results |
| **General Strategies** | Trial and error methods |
| | Simplifying difficult tasks |
| | Looking for pattern |
| | Reasoning |
| | Making and testing hypotheses |
| | Proving and disproving |
| **Personal Qualities** | Good work habits |
| | Positive attitudes to mathematics |

In the subsequent development of the British National Curriculum specification of the mathematics curriculum the division of mathematics learning objectives into facts and skills, conceptual structures, general strategies and appreciation was simplified into two categories. These are primarily mathematical content (knowledge of facts and skills and conceptual structures) and mathematical processes and personal qualities (general strategies and appreciation). This is shown in Table 2.6.

*Table 2.6: Draft Form of the National Curriculum The learning outcomes of school maths*

| COMPONENT | ELEMENTS |
|---|---|
| CONTENT | Number |
| 60% weighting | Algebra |
| | Measures |
| | Shape and Space (Geometry) |
| | Handling Data (Statistics and Probability) |
| PROCESSES | Practical Applications of Mathematics: |
| 40% weighting. | *using mathematics,* |
| | *communication skills,* |
| | *personal qualities* |

This represents the recommendations of the Mathematics Working Group, given in Dept of Education and Science (1988).

With some revision and adjustment, this dual division remained throughout the versions of the 1990s. The mathematical content in four topic areas (algebra and number having been combined) make up attainment targets Ma2 to Ma4, and mathematical processes making up attainment target Ma1. Apart from the reduced weighting of processes for assessment purposes (down to around 25%), the major change is the omission of personal qualities and appreciation, in favour of a pure processes or general strategies, Using and Applying Mathematics (attainment target Ma1).

*Table 2.7: Attainment Targets in 1995 Mathematics in the National Curriculum*

| CODE | ATTAINMENT TARGETS | COMPONENTS (KS 1-4) |
|---|---|---|
| MA1. | USING AND APPLYING MATHEMATICS | Making and monitoring decisions to solve problems<br>Communicating mathematically<br>Developing skills of mathematical reasoning |
| MA2. | NUMBER | Understanding place value and extending the number system<br>Understanding and using relationships between numbers and developing methods of computation<br>Solving numerical problems |
|  | ALGEBRA | Understanding and using functional relationships<br>Understanding and using equations and formulae |
| MA3. | SHAPE AND SPACE | Understanding and using patterns and properties of shape<br>Understanding and using properties of position, movement and transformation<br>Understanding and using measures |
| MA4. | HANDLING DATA | Classifying, collecting, representing processing and interpreting data<br>Understanding and using probability estimating and calculating the probabilities of events |

Based on Mathematics in *National Curriculum Council* (1995)

However the indications are that Using and Applying Mathematics, the first objective or attainment target is to be eliminated in favour of only mathematical content in the English, Welsh and Northern Irish National Curriculum. Following this revision in the year 2000 the emphasis will squarely be on facts, skills, concept and conceptual structures, with general strategies or creative problem solving skills downgraded.

## Evaluating The Maths Curriculum

The above discussion of the objectives of learning mathematics is best rounded out by a consideration of the assessment of mathematics learning, and the evaluation of mathematics curricula. For these are the primary functions of specifying these learning objectives.

There are three aspects of the mathematics curriculum which are distinguished in evaluation studies, e.g. Robitaille and Garden (1989), shown in Table 2.8.

*Table 2.8: The Planned, Taught And Learned Mathematics Curriculum*

| LEVEL OF MATHEMATICS CURRICULUM | *MEANS OF EVALUATING MATHEMATICS CURRICULUM* |
|---|---|
| Planned mathematics curriculum | *Curriculum outline itself is analysed and evaluated* |
| Taught mathematics curriculum: | *Curriculum observed in classroom use* |
| Learned mathematics curriculum | *Pupil's learning from curriculum assessed* |

The planned curriculum can be analysed in the ways indicated above in this chapter. The taught maths curriculum concerns classroom practice. However the learned curriculum needs to be specified in ways that reflect student performance. For this purpose, another scale has been developed by the International Assessment of Educational Progress (IAEP) project for measuring achievement in mathematics. This scale distinguishes the following levels of competence in maths (Table 2.9).

*Table 2.9: Definition of Skill Levels in Mathematics (IAEP )*

| LEVEL | SKILLS DEMONSTRATED |
|---|---|
| Level 300 | Perform simple addition and subtraction. |
| Level 400 | Use basic operations to solve simple problems. |
| Level 500 | Use intermediate level mathematics skills to solve two-step problems. |
| Level 600 | Understand measurement and geometry concepts and solve more complex problems. |
| Level 700 | Understand and apply more advanced mathematical concepts. |

This is quoted in Robitaille and Travers (1992) who report that in the international comparison of the mathematical proficiency of 13 year olds in half a dozen countries, mean scores were in the 490 to 570 point range, with USA scores lowest, South Korea highest, and UK scores around the 500 mark.

# CHAPTER 3

## LEARNING FACTS AND SKILLS IN MATHEMATICS

The above analysis of the learning outcomes of school mathematics distinguishes, first of all, facts and skills. Facts are items of information that are essentially arbitrary such as defined terms and names of concepts (e.g. odd, prime), notational devices (e.g. The decimal 'point' in place value notation, '%'), conventions (e.g. The order of operation in brackets), abbreviations (e.g. mph), conversion factors (e.g. 1 km = 5/8 m), factual results (e.g. × facts, Pythagoras' rule).

Skills are well-established multi-step procedures which involve symbolic expressions (numerical, algebraic, etc.), or geometrical figures, or the relations between them (coordinates, graphs), etc. Thus skills include performing basic operations (i.e. Employing the basic numerical algorithms), solving a linear equation, drawing a circle of given radius with compasses.

Facts must be learned primarily as individual pieces of information (although they may fit into a larger and meaningful system of facts), whereas skills are more often learned by 'doing', i.e., through repeated performance of the skilled procedures (usually on graduated examples).

A crucial factor in learning facts and skills, not surprisingly, is memory. An understanding of the distinction between three types of memory facilitates the teaching and learning of mathematics.

*Table 3.1: Memory Types*

| MEMORY TYPE | SENSORY REGISTERS | SHORT-TERM MEMORY | LONG-TERM MEMORY |
| --- | --- | --- | --- |
| **ENTRY OF INFORMATION** | Pre-attentive | Requires attention | Rehearsal |
| **CAPACITY** | Large | 7±2 items ('chunking') | No known limit |
| **MEMORY DURATION** | ¼ to 2 sec. | up to 30 sec. | Minutes to years |
| **RETRIEVAL OF INFORMATION** | Readout (like 'replay') | Automatic. Items in consciousness | Retrieval cues Search process |

The sensory registers record, without conscious attention, immediate given features of sensory experience providing "very brief persistence of information in the senses" (Riding 1977: 9). This requires no effort, and leaves a trace for up to at most 2 seconds, during which time it can be 'readout' or 'replayed'. Thus on hearing a number of noises, the memory trace can swiftly be 'replayed' and the number of sounds counted from the replay.

Short-term memory (STM) or working memory is that part of memory in which we are conscious of its contents during thinking, reasoning or imagining. It enables "retention of material during its analysis" (Riding 1977: 9). To get something into STM requires attention, like reading a mathematics task in order to do it. STM can hold around at most seven items for many people (the usual range for different persons is 7±2 items, that is 5 to 9 items). (There is a famous 1957 paper by the psychologist George Miller (1956) entitled 'the magic number 7±2' which introduced this limit). However what these items *are* varies with knowledge or experience, so to many mathematically aware persons $sinAcosB+cosAsinB$ or $\int_0^1 (y^2-7y+3)^2 dx$ each comprises at most 4 or 5 items to remember, whereas to a non mathematician, to recall all the symbols in either expression may be beyond the capacity of their STM. This is called 'chunking': the ability to combine information into larger chunks, and then treat it as a single item in memory. An elementary example is ½: this symbol is made up of 3 signs, but most of us know it as a single thing: the symbol for a half. Many children who have difficulty with something like ¼ + 1/3 would see this is a complex of seven signs, not the addition of two (fractional) numbers, which occupies 3 places in STM. Clearly chunking in this way is a vital skill that all mathematics learners need to acquire, at least up to a certain point.

STM needs to be refreshed, or its contents are replaced within about 30 seconds, as one's attention moves on.

Long term memory (LTM) has an unlimited capacity to store the memories and knowledge items, and is all the while storing information as we experience life. It comprises a "fairly permanent store of information" (Riding 1977: 9). However, to put some new symbolic information in LTM requires rehearsal: passing it through STM at least once, more often many times (like an actress learning her lines).

All learning involves storing some form or other of knowledge representation in long term memory. Some of the ways to facilitate this are described in the following section.

## THE LEARNING OF EARLY NUMBER FACTS

Research on early number learning, and in particular on the transition from counting to the use of known number facts in addition, suggests that that children's responses frequently follow a common pattern of development (Bergeron, Herscovics, and Moser 1986, Carpenter and Moser 1982, Resnick and Ford 1984). Of course, before being able to deal with the formal task of adding two numbers together young children need to have extensive experience of counting (including saying the number sequence, counting physical objects and counting repeated motifs in pictures), as well as ordering (e.g., objects and numerals by size), matching (e.g., sets of objects and pictures), and recording (e.g., numbers/numerals). Hughes (1986) has shown how the transition from informal addition of physical objects (e.g., fingers) to formal, abstract addition e.g., computing '3+2' is a source of difficulty for young children.

Putting these background issues to one side, the finding of researchers, including those listed above, is that children's responses to the task of adding two numbers together, e.g., '3+8'. frequently follows a common pattern of development with 5 stages.[2]

1. **Count all.** The child counts both numbers consecutively. 3+8 is carried out as follows (the parenthetical number and subscript show which number is being counted, and how far the count has progressed, respectively): $1(3_1)$, $2(3_2)$, $3(3_3)$, $4(8_1)$, $5(8_2)$, $6(8_3)$, $7(8_4)$, $8(8_5)$, $9(8_6)$, $10(8_7)$, $11(8_8)$. Answer 11. So the child first counts off the 3, then continues through the next 8 numbers, thus ending on the overall sum (derived by counting through all of the digits).

2. **Count on.** The child takes the first number as the starting point and counts on the number of times given by the second number: 3, $4(8_1)$, $5(8_2)$, $6(8_3)$, $7(8_4)$, $8(8_5)$, $9(8_6)$, $10(8_7)$, $11(8_8)$. Answer 11. So the child starts at 3, then continues for another 8 increments, to end on the overall sum (derived by counting on from the first number).

3. **Count on from larger.** The child commutes the sum 3+8 to 8+3 to reduce the number of steps and then counts on: 8, $9(3_1)$, $10(3_2)$, $11(3_3)$. Answer 11. So the child chooses to start at 8 (presumably having learned that 3+8 and 8+3 give the same answer), then continues for another 3 increments, to end on the overall sum.

4. **Derive from known facts** - the child uses commutative and associative transformations applied to known facts (e.g., 8+2=10, 10+1=11) as steps to the answer. Thus: e.g., one possible derivation is: 3+8 → 8+3 → 8+(2+1)→ (8+2)+1 → 10+1 → 11. Answer 11. (This example employs 'bridging through 10'.

5. **Known fact** - the child retrieves 3+8=11 directly from memory.

The learner masters these skills and facts, often in this sequence, in response to demonstrations by, or shared tasks with, more capable others in the learner's Zone of Proximal Development, i.e., that area of competence in which aid and support from more capable others is needed to complete the task (Vygotsky 1978), as well as following aided and later unaided practice. Thus for addition, the child may first have been shown (and practised) performing addition by counting all of two sets of

---

[2] In the following I am not careful to distinguish between the numeral (i.e., the number name or the sign used to indicate the number) and the number (the number concept denoted by the numeral). Although this distinction is important I assume it to be understood and a certain degree of laxity or ambiguity is not dangerous.

physical tokens (enactive representation of task, after Bruner 1960, see below), counting all of two sets of drawn tokens (iconic representation; Bruner 1960), and then counting all of two numbers (symbolic representation of task, stage 1 above). The learner child may be taught or shown that counting on shortens the procedure and is a more efficient way of completing the task, or may infer it during the experience of an extended number of addition tasks (stage 2). Likewise child may be taught, shown or induce the commutativity of addition (stage 3).

Chronometric analysis (careful measurement of the time taken to complete different tasks) suggests that children memorise certain additive number facts before others and use these (for example 0+0, 1+1, 2+2, 3+3, 4+4, ...; Resnick and Ford 1984) to derive others (stage 4). Each of these 'doubles' takes less time on average than any other 2 digit sum using the same addend. Thus typically 3+3 and 4+4 take young school children a second less to answer than 3+4, 3+5, 3+6, 4+3, 5+3, 6+3, etc., and about the same time as additions involving 1 or 0. Learning the doubles first may be a self induced strategy from extensive experience in additive tasks in school, or may taught or modelled by more knowledgeable others such as the teacher-as-researcher.

In this progression there are many experiences that contribute to the learner's developing mastery over the symbolic addition tasks. These include work with manipulatives and iconicly presented tasks, as well as oral work and a wide range of symbolic tasks solved mentally, with intermediate recording or with the electronic calculator. The symbolic tasks can include symbolic addition, subtraction (as well as multiplication and division) tasks which can be presented horizontally or vertically (in column form). In particular, the meanings that children attach to the operations of addition and subtraction are intimately related. The tasks typically also include word problems, which the child must analyse and translate into an arithmetical task before operating on any numbers. Children will have most of these experiences in the classroom, but other significant areas of activity occur in the home and other out-of-school locations. Likewise the more knowledgeable guiding other will most often be the elementary school teacher and sometimes a classroom peer, but may also include parents and others in the out-of-school locations. So it can be said that acquiring the number facts (the addition bonds) is the outcome of a complex and many-faceted process of learning and development.

## SKILLS AND ALGORITHMS

It is helpful to distinguish four different types of skills and algorithms in the domain of number: taught formal paper and pencil algorithms, informal written child methods, informal mental methods, and calculator assisted approaches.

First of all, there are the taught formal (standard) paper-and-pencil algorithms, such as column addition, division, addition of fractions:

e.g.,  132         27 ) 692        $\underline{1} + \underline{2} = \underline{5 + 6} = \underline{11}$
      +792                          3    5    15    15

Typically, these are what is taught in school.

Secondly, there are children's informal or own (i.e., self-invented) paper-and-pencil methods. The following examples have been seen to emerge spontaneously in children.

```
  2 6 3              270  --10
 -1 7 2              270  --10
  1 -1 1             540
    9 1              135  --5
                     675  Ans.: 25 r. 17
```

In the first example, to compute 263 − 172, the child subtracts the larger digit (in each column) from the smaller, giving 1,-1,1. This is 1 hundred, minus 1 ten, plus one unit, which is 91. In the second example, to compute 692 ÷ 27, the child does repeated subtraction: 10 lots of 27 off, another 10 lots of 27 off, that's 270+270=540 off; another 5 lots of 27 off is he most that can go; so 135 more off, that's 675 off in total, answer 25 remainder 17. These emerge untaught from children, and when effective (correct) usually indicate an underlying understanding of the concepts and properties involved.

Third, there are children's informal mental methods. The flexibility and choice of strategies involved are similar to problem solving (see chapters 5). These methods are discussed above and in chapters 2 and 5, and can show children's 'at-homeness' familiarity and flexibility in number operations, transformations, etc. Fischbein (1994) discusses the relationships between formal, algorithmic and children's intuitive methods in mathematics. Plunkett (1979) discusses the difference between written, formal algorithms and mental, oral, and informal methods (see Chapter 9 below).

Lastly, there are children's calculator, etc., assisted approaches in computation. These involve other aspects of skill and competence (see, e.g., Biehler *et al.* 1994).

## TEACHING FOR FACT RECALL AND SKILL CHAINING
### Rehearsal of facts and skills

The most obvious way committing facts and items to be learned to memory is by rehearsal, either by speaking what is to be learned out loud, or reading it, repeatedly. The chanting of tables falls under this heading.

The rehearsal of facts and skills is based on rote learning which is *arbitrary, disconnected,* and *verbatim.*

*Weaknesses of rehearsal of facts and skills as a sole technique.* Weaknesses include lack of transferability of knowledge and skills, and a failure of recall of any part crashes the whole procedure. Thus the whole 9× table learned by heart (rote) has to be rehearsed to retrieve one fact such as 7×9, if it is only known through rote learning. The algorithm for 3 figure column subtraction involves many decision points and procedures to recall. If learned purely by rote one slight error leads the whole procedure awry. Such an approach can lead to negative attitudes, especially following repeated failure to recall, under pressure.

*Strengths:* These include very quick retrieval of facts and skills, and the automaticity of skills – i.e., the rapid and automatic performance of algorithms and procedures without the need for conscious retrieval or reflective thought. Furthermore, it is possible, at least in part, to combine teaching/learning of skills and teaching/learning for interconnectedness. So facts and skills first learned by rehearsal (i.e., by rote and drill) can become connected with and a part of meaningful knowledge, thus overcoming the above weaknesses.

*Effective Techniques for Fact Recall.* Mnemonics help e.g. SOHCAHTOA, BODMAS. This is deliberate 'chunking', turning a relatively elaborate technical fact into a new word (or phrase) which is easier to learn by rote.

Linking to an arbitrary memorable basis is another effective technique for fact recall. For example, the British magician Paul Daniels has shown how to learn new languages swiftly by linking words to images. The famous Russian psychologist A. R. Luria (1977) describes a unique case study of a mnemonist, i.e., a man with a fantastic memory, who describes how he remembers lists of hundred of arbitrary things (objects, names, numbers) by placing them at the front doors of houses in an imagined street, or in drawers of an old-fashioned bureau, which images he can conjure up and from which he can retrieve the facts. Thus both the use of mnemonics and the linking to arbitrary memorable images can be successful ways of effective fact recall. In effect, these techniques amount to utilising or constructing a schema or conceptual structure to link the facts and aid their recall. Schemas and/or conceptual structures will be discussed at greater length in subsequent chapters. These are the major mode by which learned knowledge is stored in long term memory.

*Effective Techniques for Skill Chaining.* Skill chaining is so called because skills are made up of multi-step procedures and each step requires the recall and 'triggering' of the next step. The net result is called chaining since each step links with and calls up the next like hauling up an anchor chain. An effective technique for maintaining skill levels is distributed practice, which is based on setting learners mixed skills over a period of time. This has been demonstrated to keep knowledge of mechanical arithmetic, algebra, trigonometry skills, etc. up, better than 'block practice' time on a single skill with big gaps in between.

There are three reasons why distributed practice is better than only using 'block practice'. (of course their combination can be even more effective). First of all, distributed practice requires the skills to be retrieved less intermittently resulting in them being fresher in the mind of the learner, i.e., more recently recalled. Second, distributed practice requires learners to identify which skill is involved in a task, i.e., to select a skill from their repertoire of learned skills based on identifying the type of exercise or routine problem. Block practice can easily degenerate into learners simply repeating the same procedures that they have seen exemplified by the teacher, and beyond a certain point does not add to levels of skill retention and attainment (Reynolds and Glaser 1964). This mode of skill development does not provide the commonly needed practice in identifying which skill is needed. In a famous series of articles Brown and Küchemann (1976, 1977, 1981) report the commonly heard classroom question "Is it an add miss?" illustrating the common learner predicament in knowing a repertoire of skills but not knowing which to apply in a given task. Distributed practice is a useful technique for overcoming this. Developing an understanding of the concepts and procedures involved and acquiring problem solving strategies also provide means of overcoming this difficulty (discussed below and in a later chapter).

Third, substantial research has been carried out on this, especially in the USA, where it is a key issue in the design, evaluation and sales of mathematical text book series. Spaced-review, as distributed practice is sometimes called, has been shown in quite a number of studies to enhance learner skill retention and overall attainment on related tests. Studies by Schunert (1951), Laing (1970) and others showed significant improvements in retention and performance by students taught using distributed practice design over those only experiencing block or massed skill practice.

The Oklahoma City Public Schools adopted a DARES (Daily Assorted Review Set) programme for teaching mathematics in Oklahoma City primary schools in 1983-84. The following table shows the percentage of pupils attaining the 50th percentile score based on a standard test of mathematics achievement (primarily arithmetic).

*Table 3.2: Pupils in Oklahoma City achieving 50th percentile (standard test).*

| Year | 1980-81 | 1981-82 | 1982-83 | 1983-84 | 1984-85 |
|---|---|---|---|---|---|
| **Percent of pupils** | 45 | 46 | 46 | 52 | 55 |
| **Increase on 1980-81 score** | 0% | 2% | 2% | 16% | 22% |

Thus the scores in this table (reported in Hartzler 1988) indicate the proportion of the children who score above average marks in arithmetic. Evidently the introduction of the DARES approach has raised the proportion of the primary school population attaining average marks by a sixth (16%) or a fifth (22%). Given the size of the sample (large) this is a very significant gain. This curriculum development is reported in Hartzler (1985, 1988) who also provides the following details of the approach and claims of its effectiveness.

DARESs have these characteristics:
1) The exercises are done by students.
2) Each day's set consists of different kinds of problems, placed side-by-side, one of each kind, like a year-end achievement test.
3) A different set is done each day, though topics may repeat.
4) Each day's set may be lengthy or brief.
5) Topics are strategically sequenced from day to day.
6) Topics go back to material covered many days or weeks or months ago.
7) The problems are placed side-by-side on the blackboard, dittosheet, overhead transparency, or textbook, but never found by students turning back to previous pages in a textbook.

Benefits include the following:
1) Students keep prerequisite skills refreshed for easier learning and understanding of more advanced ideas.
2) Conceptual similarities and contrasts between topics and problem types are more available for insight learning.
3) Days are not wasted reviewing for exams.
4) Discrimination between topics is practised daily.
5) Students are much more likely to experience some success each day.
6) Less time is wasted reteaching within a given year, making time available for problem-solving, calculator, computer, etc.
7) Students may easily be directed to task at the very beginning of each period.
8) Slower students get needed reteaching and opportunity to practice.
9) Diagnosis is ongoing, especially helpful for mid-year transfer students.
10) District communication between teachers is enhanced by the writing process. Hartzler (1985: 2-3)

These details provide valuable clarification of the programme, but the claims made for its efficacy are based more on the enthusiasm of a school district supervisor than the dispassionate findings of a researcher.

Furthermore, although there is no doubt that distributed practice is an effective technique, it is also an old-fashioned technique related to a drill and practice approach and it has mainly been compared with other traditional, textbook based approaches. Nor is it unfamiliar to British primary school teachers, who have often started their lessons with assorted 'Ten a day' type exercises, similarly based on the idea of distributed practice.

At the high school level there have also been successful experiments using the distributed practice or 'incremental approach' to mathematics. Klingele and Reed (1984) compared two samples of about 300 students each on a course equivalent to

Algebra 1. Their experimental group used a textbook by John Saxon based on a distributed practice approach to algebra and their control group used a traditional algebra text employing block practice exercises. The two groups were tested using standardised mathematics achievement tests before the experiment and there were no significant differences between them.

After a semester's instruction in algebra, the scores of the two groups were as follows.

*Table 3.3: Achievement scores of algebra students*

| GROUP | Departmental Maths Exam | Basic Algebra Test |
|---|---|---|
| Control group | 42.6 | 10.9 |
| Experimental group | 52.8 | 12.1 |

Thus the experimental group scored 24% higher on the Departmental maths exam, and 22% higher on the Basic Algebra Test, thus representing significant gains, both educationally and statistically (at the 0.001 and 0.05 levels, respectively).

The experimenters were also sensitive to the criticism that the John Saxon distributed practice approach to algebra has an element of 'drill and kill' and so also tested the attitudes to mathematics of their sample using the Shatkin Mathematics Opinionnaire (a published and tested instrument for measuring of attitudes to mathematics). Their results are as follows.

*Table 3.4: Attitudes scores of algebra students*

| GROUP | Attitude inventory before instruction | Attitude inventory after instruction | Change in attitude to maths |
|---|---|---|---|
| Control group | 2.9 | 2.8 | -4% |
| Experimental group | 3.1 | 2.9 | -7% |

Although both groups' measured attitudes declined over the semester of instruction, and the Experimental group's attitude scores declined slightly more, there were no statistically significant differences between the two groups. In summary: in this experiment it can be said that the distributed practice approach to algebra was much more effective than the traditional control method in terms of achievement gains, and there were no significant statistical differences between the groups in terms of attitude measures.

Despite these positive results, not all studies have found particular implementations of the distributed practice approach to be more effective in all areas of learning. Smith (1987) compared the performance of about 140 high school students using the Saxon (distributed practice) text with about the same number using a traditional algebra text (author: Dolciani). Smith tested the students on 7 algebraic subtests. The experimental group using the distributed practice (Saxon) text scored better on 6 out of 7 subtests. However the traditional group of students scored better on the one remaining subtest on algebra definitions and theory. This is an important area and

Smith concluded that the Saxon text may have sacrificed an adequate treatment of the theoretical and conceptual areas of algebra in its focus on skills mastery.

Finally, in any skill learning the crucial role of feedback of errors to allow self-correction must not be overlooked. This will be treated in greater depth in chapters that follow.

## TEACHING FOR INTERCONNECTEDNESS

In teaching/learning for interconnectedness, facts and skills are linked to a conceptual structure. This overcomes the weaknesses of rote learning described above, but, certainly initially, lacks strengths or fast retrieval and automaticity. Ultimately, learning which is interconnected or relational, rather than rote, is more robust knowledge. (However some vividly learned rote facts can remain with the learner indefinitely.) Interconnectedness through an arbitrary basis is what was described above in terms of mnemonists, and the feats of memory experts.

R. Skemp (1976) contrasts rote learned facts and skills, which he terms 'instrumental understanding' (knowing *how to perform an* operation), with interconnected knowledge, which he terms 'relational understanding' (knowing not only *how to* but also *why*).

An effective technique for teaching facts and skills for interconnectedness, and for practising and/or reinforcing skills, is to employ and reinforce facts and skills incidentally as part of larger learning context, such as carrying out an investigational activity, or the use of a calculator for solving practical mathematical problems. Building on this idea is H. Winter's theory of practice (reported by Wittman 1985)

## *H. WINTER'S THEORY OF PRACTICE: 4 PRINCIPLES*

### PROBLEM-ORIENTED PRACTICE

Practice and repetition should relate to higher order problems. Solving these involves small steps requiring certain skills (e.g., Magic Squares). Thus learners are engaged in higher level thinking in their practice activities.

### OPERATIVE PRACTICE

Problems should be generated by systematically varying the data in order to observe patterns and get insight (e.g. reverse subtractions 83-38, 76-67, 39-93, etc.). This develops problem solving strategies.

### APPLICATION-ORIENTED PRACTICE

Wherever possible practice should be related to real situations and contribute to developing useful knowledge (e.g. ratios with food tins). This encourages meaning making, links with problem situations in the 'real' world, and can enhance motivation and the formation of positive attitudes.

### PROGRESSIVE SCHEMATIZATION

Children should first be given the opportunity to develop their own 'primitive' versions of skills and then schematise them through discussion (e.g. multiplication algorithm). Again, this encourages meaningful learning and the construction of conceptual structures.

Although these are very sensible suggestions, they are offered by Wittman and Winter as proposals only, and no systematic studies have been conducted, to my knowledge, into the efficacy of all of these principles together. A number of researchers have investigated some of these ideas individually (c.f., the work of Bruner 1960, Dienes 1960, 1970, Skemp 1976, and the Dutch realistic mathematics project, e.g., Lange 1996)

# CHAPTER 4

## CHILDREN'S ERRORS AND METHODS IN MATHEMATICS

### INTRODUCTION

It is very difficult to know exactly how children actually *learn* mathematics, since much of it is hidden in the mental functioning that lies behind their actions. There are a number of theories about how they learn but these must be, by their very natures, conjectural. One way to see what children have learned, and more important still, to see evidence of their thought processes, is to look at the errors they make in doing some mathematics. This chapter therefore focuses on children's errors in learning and doing mathematics. The richness of children's thinking that is revealed suggests something important about the way children learn mathematics.

Interesting and valuable as an examination of children's errors is, there is much, much more to their mathematical activity. For labelling something as an error assumes that there is a fixed goal which a child has failed to attain, or a procedure, which has not been used properly. Mathematics also includes problems with multiple solutions, or none, which can be approached by many methods. So the second half of this chapter looks at children solving problems or investigating mathematics. Like the first half, it also deals with children's methods, but shows that they do not have to make mistakes to show their creativity in mathematics. It also serves to remind us that learning mathematics should be a challenging activity, which excites children's curiosity and awakens their intelligence.

### CHILDREN'S ERRORS

We begin by looking at some routine mathematics, and some errors that arise. The following activity includes examples of five children's work in subtraction with whole numbers, in column form.

### ACTIVITY: DIAGNOSIS OF ROUTINE ERRORS

(1) Darren has made a mistake in some of his subtractions. What error pattern has he followed when he made his mistakes?

```
   A.        B.        C.        D.        E.        F.
  -32      -245      -524      -135      -458      -241
   16       137       298        67       372        96
  ----     -----     -----     -----     -----     -----
   16       112       374       132
```

To make sure you have found the pattern, apply Darren's method to the last two examples.

Can you describe Darren's error pattern?

..................................................................................................................................

..................................................................................................................................

..................................................................................................................................

(2) Here are three examples of Angela's work in subtraction. Can you see a pattern of errors in her work?

A.  $\begin{array}{r} {}^{8\,1}\\ 1\cancel{9}7\\ -\phantom{0}43\\ \hline 1414 \end{array}$  B.  $\begin{array}{r} {}^{6\,1}\\ 1\cancel{7}6\\ -\phantom{0}23\\ \hline 1413 \end{array}$  C.  $\begin{array}{r} {}^{7}\cancel{6}{}^{1}\\ 3\cancel{8}4\\ -\phantom{0}59\\ \hline 325 \end{array}$  D.  $\begin{array}{r} 273\\ -\phantom{0}38\\ \hline \phantom{000} \end{array}$  E.  $\begin{array}{r} 285\\ -\phantom{0}63\\ \hline \phantom{000} \end{array}$

To make sure you have found the pattern, use Angela's procedure to complete the last two examples. Now describe Angela's error pattern.

..................................................................................................................................

..................................................................................................................................

..................................................................................................................................

(3) Steve has been having problems in subtracting. Can you spot the error pattern, and apply it to D and E?

A.  $\begin{array}{r} 147\\ -\phantom{0}20\\ \hline 120 \end{array}$  B.  $\begin{array}{r} 624\\ -323\\ \hline 301 \end{array}$  C.  $\begin{array}{r} 527\\ -304\\ \hline 203 \end{array}$  D.  $\begin{array}{r} 805\\ -201\\ \hline 604 \end{array}$  E.  $\begin{array}{r} 446\\ -302\\ \hline \phantom{000} \end{array}$  F.  $\begin{array}{r} 760\\ -230\\ \hline \phantom{000} \end{array}$

Describe Steve's error pattern.

..................................................................................................................................

..................................................................................................................................

..................................................................................................................................

(4) Here are some examples of Donna's work. Can you see an error pattern?

A.  $\begin{array}{r} {}^{8\,1}\\ \cancel{6}\cancel{9}3\\ -248\\ \hline 445 \end{array}$  B.  $\begin{array}{r} {}^{2}\cancel{3}25\\ -151\\ \hline 174 \end{array}$  C.  $\begin{array}{r} {}^{5}\cancel{7}{}^{1}26\\ -349\\ \hline 287 \end{array}$  D.  $\begin{array}{r} {}^{2}\cancel{3}4\\ -276\\ \hline 68 \end{array}$  E.  $\begin{array}{r} 436\\ -172\\ \hline \phantom{000} \end{array}$  F.  $\begin{array}{r} 625\\ -348\\ \hline \phantom{000} \end{array}$

Apply Donna's method to the last two questions. Can you describe her error?

..................................................................................................................................

..................................................................................................................................

..................................................................................................................................

(5) Tim's method of subtraction seems impossible to analyse, or is it?

A. $\overset{2}{\cancel{3}}\overset{4}{\cancel{5}}$   B. $\overset{3}{\cancel{3}}\overset{3}{\cancel{4}}0$   C. $\overset{}{\cancel{3}}\overset{6}{\cancel{4}}3$   D. $\overset{}{\cancel{5}}\overset{6}{\cancel{7}}0$   E. $385$   F. $640$
  $-21$       $-205$       $-341$       $-443$       $-322$       $-626$
  $\overline{\phantom{0}13}$   $\overline{130}$   $\overline{112}$   $\overline{120}$

If you can spot an error pattern, apply it to E & F, and then describe it.

.................................................................................................................

.................................................................................................................

.................................................................................................................

## DISCUSSION OF THE CHILDREN'S METHODS

(1) In B, C and D Darren seems to be taking the smaller digit from the larger, in each column, irrespective of whether it is on the top or bottom line. You have probably seen this error in the classroom. It indicates confidence in single digit subtraction, but a failure to have grasped multi-digit subtraction procedures. Using his method Darren would get (E)=126 and (F)=255. His answer to (A) is an interesting exception. Probably the smaller numbers made sense to him, and he used some fact he knew, such as 32-16=16, or 16+16=32.

(2) Angela seems to be 'borrowing' (decomposing one of the tens into units) whether she needs to or not. So she has learned the decomposition method of subtraction, except she does not know quite when to apply it. Her method can give a two digit answer in the units column, and when she does this (e.g. in A: 17-3=14) it throws her answer way off. This suggests that she has not fully grasped the concept of place value, since her two digits of Units have also occupied the Tens place. Her method gives (D)=235 and (E)=2112.

(3) Steve operates correctly within the columns of subtractions without regrouping (decomposition), except when a zero is involved. He gets 7-0=0 in A, and 2-0=0 in C. He seems to be mixing up the roles of zero in multiplication and addition (7×0=0 and 2×0=0). This is not all that surprising in view of the special role zero plays with both operations. Steve seems to be remembering these special roles, but confusing them. He would get (E)=104 and (F)=530.

(4) Donna can 'borrow' successfully from either the tens or hundreds columns. But when she needs to 'borrow' from both, she decomposes one Hundred into ten Tens (correctly) and one into ten Units (incorrectly). It looks like she thinks she cannot 'borrow' one of the Tens, when they are exceeded by the number of Tens to be taken. Her method gives (E)=264 and (F)=187.

(5) Tim's method can take a while to spot. It appears that he works his subtractions from left to right. Like Angela, he borrows whether he needs to or not (but 'decomposes' a Unit into ten Tens, or a Ten into ten Hundreds). When he comes to his final column (the Units) he does the best he can, and puts zero for any negative answers (0-5=0 in B, 0-3=0 in D). Because of the direction in which he works his subtractions Tim cannot increase his number of units by decomposition. Tim's method gives (E)=153 and (F)=110.

**WHAT THE ERRORS MEAN**

These examples of children's actual errors suggest a number of things about their thought processes. First of all, it is clear that the errors are systematic, the results of partially correct procedures for subtraction being used more or less consistently. There is no question of the errors being random or being due to carelessness, in view of the undeniable patterns in them. This corresponds to researchers' observations about the systematicity of error patterns (Ashlock 1976, Ginsburg 1977, Lankford 1972, Resnick and Ford 1984).

Secondly, it seems likely that the children have constructed their own faulty subtraction procedures, presumably based on correct demonstrations. How can this happen? Evidently the children have not simply stored in their memories the procedure they have been taught (Angela, Donna and Tim show evidence of being taught the decomposition method). Rather they seem to have absorbed some of the different components, and have reassembled them in their own individual ways. In other words, the children have recreated their knowledge (in this case, of the subtraction algorithm) in their own unique, original and personal ways. The originality of this constructed knowledge is evident from it's incorrectness. It is absurd to think that these children have been taught the weird yet systematic procedures they use. This suggests very strongly that they must have created it themselves.

These examples, and the discussion of them that follows, illustrate the view of learning known as Constructivism: Children are the active constructors of their own knowledge, albeit on the basis of the sense they make of their experiences, classroom or otherwise, and drawing on their previous knowledge.

A classic early case study of one child's construction of his own interpretation of mathematics is given by the US researcher Stanley Erlwanger (1973), who interviewed 'Benny', a boy who had successfully completed an individualised scheme called *IPI Mathematics* (Individually Prescribed Instruction). Erlwanger uncovered idiosyncratic and erroneous rules derived by Benny from his learning, as well as idiosyncratic conceptions of rules, answers and of mathematics itself. Benny's conception of mathematics results directly from his experience of learning mathematics through the individualised IPI mathematics scheme. Although much of the personal knowledge Benny constructed was incorrect and distorted, he still scored 60-70% on the assessments built into the scheme, and his problems were not picked up by the classroom teacher or the teaching assistant. Many children in Great Britain also learn mathematics through individualised schemes. To what extent can their experiences be like Benny's?

## LOOKING AT ONE ERROR

One of the erroneous rules invented by Benny is: 'When adding fractions, add the top numbers and put the result over the sum of the bottom numbers.' This is the Adding Numerators and Denominators (*AND*) error: $\frac{a}{b} + \frac{c}{d} = \frac{a+c}{b+d}$. This is a well known error which you have probably seen before in the classroom. Kath Hart (1981) reports that many children commit this error. The following table illustrates some of the results from that study:

*Table 4.1: Pupil responses on Fraction Tasks*

| TASK TYPE / CHILD'S RESPONSE | AGE 12 | AGE 13 | AGE 14 | AGE 15 |
|---|---|---|---|---|
| **FORMAL COMPUTATION** | | | | |
| Error: 1/3 + 1/4 = 2/7 | 18% | 29% | 22% | 20% |
| Error: 3/8 + 2/8 = 5/16 | 9% | 20% | 14% | 17% |
| Correct: 3/8 + 2/8 = 5/8 | 78% | 66% | 72% | 68% |
| **PROBLEM VERSION:** ...*3/8 of flour used for bread and 2/8 for cakes, etc.* | | | | |
| Error: 3/8 + 2/8 = 5/16 | 5% | 7% | 7% | 6% |
| Correct: 3/8 + 2/8 = 5/8 | 77% | 73% | 81% | 77% |

These results have some surprising implications. First of all, despite all of the years of primary schooling and the fraction teaching there, a surprisingly large number of children committed the error in question (adding denominators). Between 10% and 30% of all of the children tested (approximately 1200 children were tested on fractions, in total) committed this error in routine fraction computations.

Secondly, when the addition of fractions appears in a more meaningful context (such as in the flour problem) it seems that the children are less likely to seek mechanical procedures or rules to solve the problem. They seem more likely to see the problem in terms of everyday experiences, and to use some form of informal reasoning to solve the problem. (for example, their thinking may be supported by mental images or pencil sketches of 3/8 of a bag of flour and 2/8 of a bag of flour; whatever they are imagining it is probably not just symbols.) Thus, the 'adding denominators' error appears less often (only 5% to 7%), as opposed to 9% to 20% when the same problem is presented as a formal computation.

The Assessment of Performance Unit (1985) tested an even larger sample of children, and found the same AND error to be by far and away the most common error in adding fractions. They were able to describe the occurrence of this error in terms of the kind of fraction addition task as follows:

- with the same denominators - up to 10% committed the error at 11 and 15 years.
- with different but familiar denominators (½'s, ¼'s, 1/8's) - up to 15%, at age 11, and 10% at age 15 committed the error.
- with different and unfamiliar denominators - up to 25% at age 11, and 15% at age 15 committed the error. (Assessment of Performance Unit 1985, page 132.)

This perhaps shows that the further children get from simple and familiar cases, the more difficult they find it to recall and apply the appropriate rule for adding fractions.

Overall, an examination of one particular error suggests that rules for computation with fractions are not very well remembered, especially as the rules and situations for applying them get more complex, and that children will readily invent a rule or procedure, if they need one, just as Benny does. What is perhaps most surprising is the sheer number of children who invent for themselves the same false AND rule, up to 25% of 11 year olds!. One must speculate that it arises from (a) lack of understanding of the processes of fraction addition, coupled with interference from the easier multiplication rule $\frac{a}{b} \times \frac{c}{d} = \frac{ac}{bd}$. An interesting final note about the AND error and children's level of attainment in mathematics is as follows. The APU found that the error is rare among the above average attaining students, increases significantly in the middle band, and is generally the most popular response for the below average attainers.

## ACTIVITY: DIAGNOSIS OF NON-ROUTINE ERRORS

The following activity illustrates two non-routine mathematical problems and a number of children's answers to them. Solve the problem yourself and them examine the children's solutions to see if you can see how they solved the problems and what went wrong, if anything.

# PROBLEM 1

*You can choose Vanilla, Chocolate, Strawberry, Lychee, or Tutti-frutti flavoured ice cream. How many different two-flavour cones are possible.*

Three 11 10 year old's answers were as follows. (Right answer 10). Identify their errors, if any, and conjecture what reasoning processes they used.

---

A.

VC VS VT VL TL
CS CL CT CV SL ST

ans = 11

---

B.

*(pentagon diagram with vertices V, C, S, L, T connected by diagonals)*

ans = 9

---

C.

   5        +      4
first choice   second ch.

ans = 9

# PROBLEM 2

*Six children hold a chess tournament. Each 2 children play just one game. How many games are played altogether?*

Four 12 year old's answers to this question are shown. (Right answer 15). Identify their errors, if any, and conjecture what reasoning processes they used.

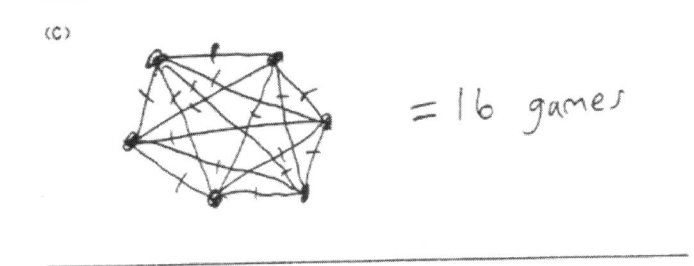

---

(A) $6 \times 5$  ( cos he can't play himself )
　　　= 30 games

---

(B)  
$\begin{array}{r} 6 \\ 5 \\ 4 \\ 3 \\ 2 \\ 1 \end{array}$ = 21 games

---

(C) = 16 games

---

(D)　15 games

## CHILDREN'S ERRORS

We have seen a few samples of children's work and evidence of how children think in mathematics. In particular we have seen many examples of children's errors with whole number operations and fractions. These show that far from being random or careless, the errors result from incompletely developed concepts (such as the concepts/conceptual structures of place value and fraction) or incorrect procedures (such as column subtraction and fraction addition), or incorrect representations or processing of non-routine problems. These errors provide plausible evidence that children construct their knowledge, be it conceptual knowledge or procedural skills, over a period of time *in their own, unique ways*. In this process of construction they may fill in any gaps in their learning with additional knowledge of their own creation. Thus errors provide a testament to children's resourcefulness and creativity, as well as an insight into how children think.

The insight provided is that children construct their own meanings and understandings, on the basis of the multiplicity of experiences and social interactions they have, including those at school, based on their previous knowledge. This is the view of learning, now widely accepted, known as Constructivism.

The evidence for children's construction of their own knowledge and understanding is there for any parent to see. Babies are born with virtually nothing in the way of knowledge, except for the ability to learn. With this they create their own understanding of their mother tongue. Clearly they must create and build their own concepts and knowledge. Similarly children create their own mathematical concepts and procedures themselves, as a part of learning to speak, to understand and communicate with others, and to identify and operate on things in the world of their experience. As in the case of their mother tongue, children have to build up (construct) their own mathematical ideas and concepts. The important point here is that mathematical knowledge cannot be given or directly transmitted to children. All that we can do as teachers (and before that, as parents) is to offer children opportunities to construct their own understandings of mathematics. Where their mental constructs do not fit with the concepts and procedures intended by the teacher, for whatever reason, the errors that we have been considering become evident.

It should be emphasised again how essential errors are to the learning process. The learning of any concept requires the boundary of the concept to be defined in terms of the contrast between what it applies to and what it does not. Attempting to apply a concept to a non-example, which this process of definition requires, is called an error, but is a necessary part of concept formation. Likewise trial-and-error (or trial and improve, as it is sometimes called) is an essential problem solving strategy which, by its very nature, requires errors. So errors are an essential part of learning. It is not possible to eliminate the unique, personal nature of any learner's knowledge, which lies at the heart of errors in mathematics. Successful teaching can only ensure that the fit between the learner's knowledge and official mathematical knowledge is good enough to result in the accepted outcomes.

It is worth remarking that despite the necessity of errors in the learning process, errors are often castigated and thought to be bad. Synonyms of 'error' are failure, bad result, blunder, fault, offence, transgression, wrongdoing, to err, to do wrong, to offend, even to commit a sin. Children are often ashamed to make mistakes and will often tear out the pages of their exercise books to start again without mistakes. This is a terrible and anti-educational thing. It is to mix up the importance of the ends and means of learning mathematics. Children's written work is just a means to an end, and that end is the learning of mathematics. When exercises have been completed and an exercise book is full, it has little educational value, and can often be thrown away, for all the use it provides the Average learner. What counts is what they take away with them; the knowledge, skills and capabilities that they can build on and apply in future learning and life activities.

Part of the 'anti-error' ideology, which condemns any errors children make, is based on a false view of mathematics, namely that mathematics is all about right versus wrong answers and that everything is exact and certain in mathematics. Such a view can be associated with less confident teachers (although it need not be). In addition, teachers often worry that when children make mistakes or errors, if they are not pointed out and corrected immediately, then they will become reinforced and etched ever more permanently into the mind of the learner. While errors do ultimately need correction, preferably as soon as possible if issues of confidence and attitude are not at stake, a negative attitude to them by teachers or others-researcher is quickly picked up by children who can then become ashamed, try to hide the errors, and subsequently avoid risk-taking in mathematics. So an attitude has to be developed whereby both children and teachers admit to their errors in public, without shame or concealment, as otherwise creative problem solving, which involves risk taking (i.e., taking the risk of making errors) cannot occur. The importance of such approaches to errors, indeed the necessary role of errors in learning and coming to know, are widely recognised in the literature (Askew and Wiliam 1995, Novak 1987).

## INVESTIGATING CHILDREN'S ERRORS IN MATHEMATICS

Children's errors in mathematics are probably best uncovered by what is called a clinical interview, in which the child works through some tasks (often a carefully graded sequence), and the interviewer asks the child to explain the methods used, and probes to uncover the underlying conceptions. This is the approach adopted by constructivist researchers, since Piaget's pioneering work.

In carrying out such a procedure it is helpful to record the interview (video or audio) which together with the child's written work, allows for a full analysis afterwards. It is possible to take notes instead of recording, but this can distract both of you. Also, it really is much better to have the child's actual words to analyse.

A key feature of good clinical interviewing is the development or use of tasks for the child or children to work through which utilise multiple modes of representation. If investigating children's numerical skills tasks can be presented (1) as formal sums, (2) as word problems, (3) orally, (4) with practical teaching aids or concrete representations of number, (5) on a computer or with a calculator, and (6) with diagrams, etc. By investigating children's responses to a variety of task representations one can gain deeper insights into their conceptual understandings and into their underlying competence in the topic area, and not just their achievement level on one type of tasks. Such an approach also helps to uncover errors, misconceptions, and imbalances between their capacities on different task types.

In addition to or together with the clinical interview approach, it is possible to use a diagnostic test to uncover children's understanding. Many tests are published, by NFER for example, including those developed for the research in Hart (1981), the *Chelsea Diagnostic Mathematics Tests*, NFER-Nelson, 1985, for 11-16 year olds (Hart 1986). Now out of print but available through the Shell Mathematics Centre, School of Education, Nottingham University. (A copyright-free version is given in Hart 1980).

A word of warning: many of the tests published as 'diagnostic' would be better described as achievement tests. The more topics such a test covers, the less it tells you about children's understanding in any fine grained way. A reference on diagnosis and diagnostic tests is Underhill *et al.* (1980).

There are a large number of publications reporting research on children's methods, errors and personal conceptions in mathematics. Different examples of children's methods, errors and understandings are reported in Ashlock (1976), Assessment of Performance Unit (1979-1981, 1985, 1991), Booth (1982), Brown and Küchemann (1976-81), Dickson *et al.* (1984), Ginsburg (1977), Hart (1978) Hughes (1986), Kent (1978, 1979), Lankford (1972), Larcombe (1985), Rees and Barr (1984), Resnick and Ford (1984), Thyne (1954), and Ward (1979).

*Table 4.2: Characterisation of sample subtraction 'bugs' (from Brown and Van Lehn 1982)*

| | |
|---|---|
| 1) <u>Always-Borrow-Left</u> The student borrows from the leftmost digit instead of borrowing from the digit immediately to the left.<br>E.g.    733<br>        -216<br>         427 | 8) <u>Smaller-From-Larger-Instead-Of-Borrow-From-Zero</u> The student does not borrow across a 0. Instead he or she will subtract the smaller from the larger digit.<br>E.g.   306      306<br>      -  8     -148<br>      302      162 |
| 2) <u>Blank-Instead-Of-Borrow</u> When a borrow is needed the student simply skips the column and goes on to the next.<br>      425<br>    -283<br>       22 | 9) <u>Stops-Borrow-At-Zero</u> Instead of borrowing across a 0, the student adds 10 to the column he or she is doing but doesn't decrement from a column to the left.<br>E.g.   404<br>     - 22<br>     227 |
| 3) <u>Blank-Instead-Of-Borrow-From-Zero</u> When a borrow from zero is needed the student skips the column and goes on to the next.<br>E.g.   507<br>     -241<br>       33 | 10) <u>Stutter-Subtract</u> When there are blanks in the bottom number, the student subtracts the leftmost digit of the bottom number in every column that has a blank.<br>E.g.   4369<br>     -  22<br>     2147 |
| 4) <u>Borrow-No-Decrement-Except-Last</u> Decrements only in last column of the problem.<br>E.g.    6262<br>     -4444<br>     1828 | 11) <u>Zero-After-Borrow</u> When a column requires a borrow, the student decrements correctly, but writes 0 as the answer.<br>E.g.   65<br>    -48<br>     10 |
| 5) <u>Borrow-Won't-Recourse</u> Instead of borrowing across a 0, the student stops doing the exercise.<br>E.g.    8035<br>    -2662<br>        3 | 12) <u>Zero-Instead-Of-Borrow</u> The student doesn't borrow, he or she writes 0 as the answer instead.<br>E.g.   42<br>    -16<br>     30 |
| 6) <u>Doesn't-Borrow</u> The student stops doing the exercise when a borrow is required.<br>E.g.   833<br>    -262<br>       1 | 13) <u>Zero-Instead-Of-Borrow-From-Zero</u> The student won't borrow if he or she has to borrow across 0. Instead he or she will write 0 as the answer to the column requiring the borrow.<br>E.g.   702     702<br>     -  8    -348<br>     700     360 |
| 7) <u>Don't-Decrement-Zero</u> When borrowing across a 0, the student changes the 0 to 10 instead of a 9.<br>E.g.   506<br>    -318<br>     198 | |

More theoretical accounts of children's errors and faulty algorithms and procedures, including computer-modelling of errors, are given in Attisha and Yazdani (1984), Bright (1984), Carpenter *et al.* (1982) and Whetton and Hagues (1985). Brown and Burton (1978) describe diagnostic models for procedural bugs in basic mathematical skills based on computer programmes. Brown and Van Lehn (1982) and Van Lehn

(1983) proposes a cognitive science-based model of how error patterns are created and repaired in learning procedures, which they term 'repair theory'. Brown and Van Lehn (1982) use this theory to describe 13 common error patterns in whole number subtraction. (See Table 4.2 above). Barclay (1980) even describes an activity for children in which they have to spot the mistakes in computations.

## TEACHING TO OVERCOME ERRORS

Booth (1985), Hart (1985) and Kerslake (1985) describe experimental teaching programmes in a research project called Strategies and Errors in Secondary Mathematics (SESM). In this project, following the identification of common errors in algebra, ratio and fractions (respectively) teaching programmes built on anticipating and correcting these errors were designed, tested and evaluated with school children. Generally the outcomes were favourable.

A variant of the method employed in these studies is the discussed diagnostic teaching approach which has been developed by Alan Bell and colleagues in a number of publications (e.g., Bell 1989). The principles of this approach are as follows:

1. Identify the key concepts and the main misconceptions, or deep-seated errors, in a particular mathematical topic through research (or the research literature).
2. Devise activities that elicit problematic or erroneous responses from students who have these misconceptions and deep-seated errors, and make sure that these answers will tend to deviate visibly from the conventionally 'correct' answers in the topic.
3. Arrange the class in co-operative groups with the instruction to discuss the answers between themselves and come up with a shared or agreed answer.
4. Get the groups of students to present their answers to the class, and get the whole class to discuss the answers. (The teacher plays the role of impartial chair and accepts all answers and gives no indication of or hints about correctness for answers during the process).

The aim of this approach is, given that the chief conceptual obstacles in a topic have been identified, to provoke cognitive conflict amongst the students by having them confront a variety of answers to the tasks and forcing them to work out, argue for and defend in discussion one of these answers. When presented with conflicting explanations, accounts and answers, learners will usually consciously and unconsciously attempt to resolve the conflict. Such situations can be very successful in provoking changes in their conceptual understandings, and restructuring of their conceptual structures and knowledge.

Bell (1989) describes the diagnostic teaching approach and provides evidence of its efficacy by reporting on a number of research studies. Bell's diagnostic teaching approach has proved remarkably efficacious working with experimental groups, who not only gained considerably in terms of test performance and understanding of the mathematical topics involved, but whose gains continued to increase as measured by delayed post-tests.

Other researchers have developed cognitive conflict approaches as the kernel of constructivist-based (see the following two chapters) teaching and learning programmes, beginning with errors, child methods, learner conceptions, misconceptions or alternative frameworks. Driver and Oldham (1986) designed a constructivist teaching sequence for science education based on these ideas, which is shown in Figure 4.1.

*Figure 4.1: Constructivist teaching sequence based on cognitive conflict*

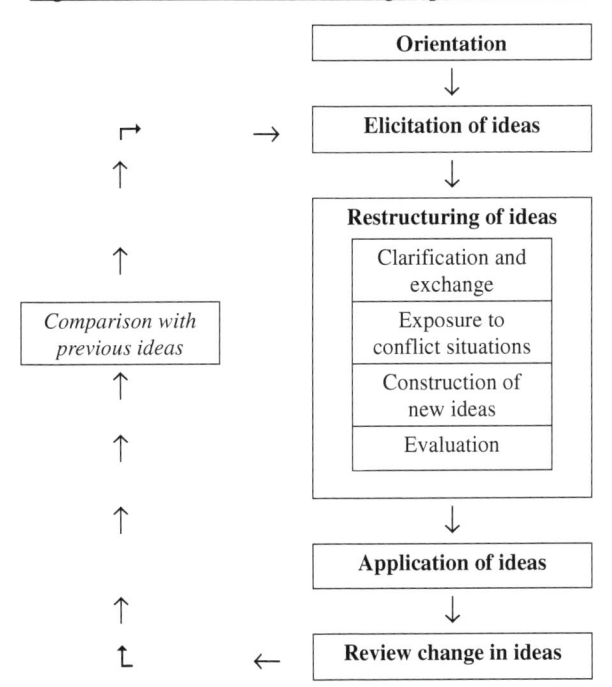

```
                    ┌─────────────────────┐
                    │    Orientation      │
                    └─────────┬───────────┘
                              ↓
    ┌→      →        ┌─────────────────────┐
    ↑                │ Elicitation of ideas│
                     └─────────┬───────────┘
    ↑                          ↓
                    ┌─────────────────────┐
    ↑               │ Restructuring of ideas │
                    │  ┌────────────────┐ │
 ┌──────────┐       │  │ Clarification and│ │
 │Comparison with│  │  │    exchange      │ │
 │previous ideas │  │  ├────────────────┤ │
 └──────────┘       │  │ Exposure to      │ │
    ↑               │  │ conflict situations│ │
                    │  ├────────────────┤ │
    ↑               │  │ Construction of  │ │
                    │  │  new ideas       │ │
    ↑               │  ├────────────────┤ │
                    │  │  Evaluation      │ │
    ↑               │  └────────────────┘ │
                    └─────────┬───────────┘
    ↑                         ↓
                    ┌─────────────────────┐
    ↑               │ Application of ideas│
                    └─────────┬───────────┘
    L      ←                  ↓
                    ┌─────────────────────┐
                    │Review change in ideas│
                    └─────────────────────┘
```

The teaching sequence comprises five phases: Orientation, Elicitation, Restructuring, Application, and Review. The sequence begins with an orientation phase in which pupils have the opportunity to develop a sense of purpose and motivation for learning the topic. In the Elicitation phase pupils make their own ideas and methods explicit, bringing them to conscious awareness. This can be by discussion, designing posters, writing, etc. The next phases is that of restructuring, which includes a number of different aspects. Once pupils' ideas and methods are out in the open, clarification and exchange occurs through discussion. Once the meanings and methods are 'sharpened up', alternative views may emerge spontaneously or may be pointed out. Alternatively surprise demonstrations by the teacher can reveal inconsistencies through disconfirming examples, thus deliberately promoting exposure to conflict situations. In response, pupils' need to construct or reconstruct new ideas to accommodate and overcome the conflicts. Pupils also need to evaluate the new ideas, thus finishing off the restructuring of ideas phase (where successful).

In the application phase, pupils use their developed of ideas and methods in both familiar and novel situations to consolidate and reinforce them. In the final review phase, pupils are invited to reflect back on how their ideas have changed, comparing their new ideas and methods to those they started with, and possibly writing reflective journals. In Figure 4.1 this is shown in the backward loop connecting the last and second phases.

This sequence strongly matches the diagnostic teaching method employed by Alan Bell and colleagues described above.

# CHAPTER 5
## CHILDREN'S LEARNING OF CONCEPTS AND STRUCTURES

Children's errors and misconceptions, as well as their strategies and creative thinking, show that children are not simply vessels which fill up with knowledge transmitted by their teachers. Rather children construct their own knowledge over a period of time *in their own, unique ways*. They construct their own meanings and understandings on the basis of the multiplicity of experiences and social interactions they have, including but not restricted to those they have at school. Not surprisingly, children's construction of their own knowledge often gives results other than those intended by the teacher. When a specific piece of knowledge is intended, we term this idiosyncratic knowledge or procedure of the child's an error or misconception. In other circumstances, such as in solving a problem, we regard the unanticipated associations as providing original strategies and creative thought. This view of learning, in which children mentally construct their own knowledge, is known as constructivism.

---
ACTIVITY

Calculate 112 - 89 mentally, taking note of how you do it.

---

Most people use a variety of informal mental methods, such as can be recorded as the following five (of many, many possible) approaches:

89, 99, 109, 110, 111, 112....= 23 (count up to the next decade then count on in decades, and then add on the remaining unit steps);

89, 99, 109, 112....= 23 (similar but counting on in decades from the starting number);

89 to 100 = 11, 100 to 112 = 12, 11+12=23;

112 to 100 = 12, 100 to 89 = 11, 12+11=23;

89, 90, 100, 110, 112 = 1+10+10+2 = 23, etc.;

(as well as the usual algorithm: place-value column subtraction).

The five informal methods shown illustrate how even in carrying out a simple subtraction we use our own methods and strategies. These methods, with the exception of the vertical subtraction, have probably never been taught to us. In other words, most of us create our own, original methods and strategies. This is shown in a particularly vivid way in cases of calculating prodigies and in cases of calculators with brain damage, who can evolve shortcuts and other means of truncating or substituting unusual new methods in their calculations.

TABLE 5.1

Skills included in descriptive framework, listed by facility (hardest first)

| | |
|---|---|
| 3 | Mentally carried out two digit *'take away'* with regrouping |
| 6 | Uses multiplication facts to solve a *'sharing'* word problem |
| 47 | Perceives *'compare difference unknown'* word problem as subtraction |
| 4 | Models two digit *'take away'* with regrouping apparatus |
| 45 | Fully appreciates concept of class inclusion |
| 7 | Mentally carries out two digit take away without regrouping |
| 5 | Uses multiplication facts to solve a *'lots of'* word problem |
| 2 | Mentally carries out two digit addition with regrouping |
| 20 | Uses counting back/up/down strategy for *'take away'* |
| 33 | Groups objects to enumerate a partly grouped collection |
| 15 | Uses repeated addition/subtraction for a *'sharing'* word problem |
| 46 | Partly appreciates concept of class inclusion |
| 12 | Uses derived facts for addition |
| 8 | Mentally carries out two digit addition without regrouping |
| 24 | Can count in 10s from a non-decade two digit number |
| 25 | Can count backwards in 10s from a non-decade two digit number |
| 34 | Quantitatively compares differently grouped collection |
| 9 | *'Knows answer'* when taking ten away from a two digit number |
| 10 | *'Knows answer'* when adding ten to a two digit number |
| 1 | Models two digit addition with regrouping using apparatus |
| 17 | Knows number bonds (not just the 'doubles') |
| 26 | Interpolates between decade numbers on a number line |
| 13 | Models two digit *'take away'* (no regrouping) using apparatus |
| 16 | Uses repeated addition for a *'lots of'* word problem |
| 19 | Solves *'compare (more) difference unknown'* word problem |
| 21 | Counts in 2s and 1s for a collection grouped in 2s |
| 11 | *'knows answer'* when adding units to a decade number |
| 14 | Models two digit addition (no regrouping) using apparatus |
| 22 | Counts in 5s and 1s for a collection grouped in 5s |
| 18 | Solves *'compare (more) compared set unknown'* word problem |
| 23 | Counts in 10s and 1s for a collection grouped in 10s |
| 40 | Knows numbers backwards from 20 |
| 27 | Orders a selection of non-sequential two digit numerals |
| 35 | Appreciates structure of grouped collections |
| 38 | Solves sharing problem by direct physical modelling |
| 39 | Solves *'lots of'* problem by direct physical modelling |
| 44 | Appreciates conservation of number |
| 28 | Appreciates commutativity of addition for *'1+n'* |
| 29 | Uses a counting-on strategy for addition |
| 30 | Reads a selection of non-sequential two digit numerals |
| 32 | Repeats numbers in correct sequence for counting in 2s 5s and 10s |
| 37 | Uses counting on strategy when provoked |
| 31 | Repeats numbers in correct sequence to 99 |
| 41 | Knows numbers backwards from 10 |
| 36 | Compares collections and states whether equal |
| 42 | Can say numbers in correct sequence to 20, can solve addition and take away by direct physical modelling |
| 43 | Makes 1:1 correspondence |

## EXAMPLE: MAPPING CHILDREN'S NUMBER CONCEPTS

Brenda Denvir and Margaret Brown at King's College, Chelsea carried out a research project in which children's early number concepts from counting to two digit subtraction with regrouping (i.e., 'carrying' or regrouping) were analysed, and a learning hierarchy constructed (Denvir and Brown 1986a, 1986b). This shows the order in which the different concepts and skills need to be learned, in theory, because of the way that some concepts and skills depend on others. The different components of this analysis and the way they relate in the hierarchy are shown below.

Table 5.1 lists all the skills that the researchers decided were necessary for a child to carry out two digit subtraction with regrouping (that is with borrowing or decomposing one 10 into ten 1's).

What the researchers next did was to make a hierarchical framework of the skills listed in the table, with arrows denoting prerequisite skills, and dotted lines showing strong connections that hold between different skills. Figure 5.1 shows the complex structure of number and skills that the researchers have hypothesised. Simpler skills are shown lower down the network, and the arrows show the presumed learning order. This hierarchy is arrived at by a logical analysis, and children do not have to fit in with it, as we shall see.

Figure 5.1 shows one subject's performance at the pre-tests, post-tests, and delayed post-tests on many of the skills. It illustrates the number knowledge of this one child (identified as 'Jy') both before and after teaching. What is fascinating and astonishing is that although Jy has been taught 8 of the concepts or skills (and only learnt 4 of these), Jy has also learnt 5 of the untaught concepts or skills. This illustrates the point made above: how much what is learnt depends on the child. Teaching may help, but ultimately it is the child who constructs his or her own mathematical concepts and skills; in short, who constructs his or her own understanding and learning.

Unfortunately it has not been possible to label the nodes in Figure 5.1 and the interested reader is referred to the original paper (Denvir and Brown 1986a, 1986b).

It should be mentioned that this project was concerned with slow learning primary school children. However the principles of learning that are suggested by it are just as applicable to children of average and above average attainment in mathematics, who should be at least as good at constructing their own understanding of mathematics.

# FIG 5.1: CHART FOR ONE CHILD (Jy)

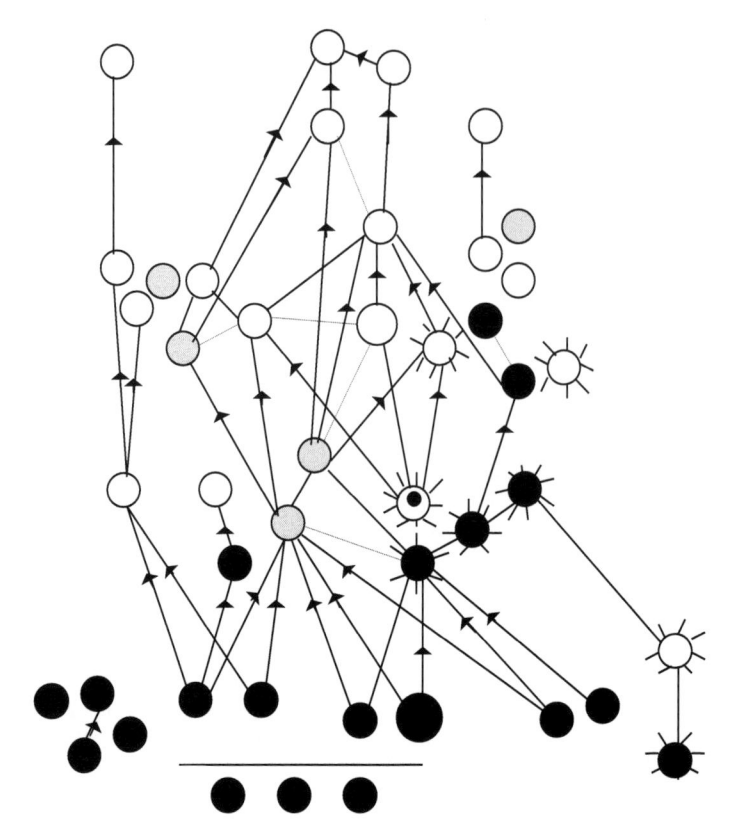

- ● indicates skill acquired at pre-test
- ◯ (grey) indicates skill acquired at post-test
- ⊙ indicates skill acquired at delayed post-test
- ○ indicates skill not acquired
- ☼ indicates taught skill

*Fig 5.1 Jy's performance at pre-test, post-test and delayed post-test in relation to the hierarchical framework.*

# CONSTRUCTIVISM

In brief, constructivism is the theory of learning according to which a child actively constructs his or her understanding on the basis of his or her experiences, including intellectual and linguistic experiences, physical activities and practice, and the child's emotional experiences, using previous understandings to make sense of these experiences. Note that the active constructing of understanding can be a purely mental activity, with no physical or other visible signs of activity. The activity is a mental one as the child links new ideas and learning with existing conceptual structures. So to identify the activity necessary according to the constructivist theory of learning with discovery learning or problem solving is a misunderstanding.

The constructivist theory of learning, in its general form, is now very widely accepted. In the psychology of learning mathematics it derives primarily from the work of Jean Piaget, although similar insights can also be found in sociology (e.g., Berger and Luckmann 1966). Piaget's theory of mental development and its stages is no longer fully accepted. Research on children's learning has raised serious doubts about his Stage theory, and his notion of readiness for learning. Although the theory is no longer accepted in its entirety, many of his ideas are still widely accepted. However two of Piaget's central insights remain valid. The first is that children learn from their interaction with the world, both from their social interactions (in which language is central), and from their experiences of the physical world, including handling objects (which leads them to make mental representations, which they are then able to transform mentally). It is now realised that the role of language is of great importance, and requires more emphasis than Piaget gave it (Donaldson 1978, Vygotsky 1986).

The second insight concerns the ways in which children's knowledge grows. Biology presents us with the view of organisms who all the while must adapt to survive. Similarly, we are all the time naturally adapting our knowledge structures, our mental maps of the world including the schematic guides for our actions, in order to survive, according to Piaget. As children (and indeed all persons) experience a new situation they try to make sense of the situation, by interpreting it in terms of their existing mental schemas. (A schema is an organised unit of information, a cluster of interrelated concepts, a conceptual structure.) This process of organising and storing new information by means of a schema is called assimilation. Learning by assimilation is incremental: the informational traces of experience are added to schemas by increments. Thus, for example, children will assimilate their learning of three digit subtraction to their schema for two digit subtraction, in a number of increments.

When a new experience cannot be interpreted in terms of the existing schemas, a (cognitive) conflict is caused. This conflict is resolved by changes in the schemas, a restructuring or accommodation of the schemas, or the creation of a new schema. (In extreme cases it can lead to stress and mental breakdown, but this is more common among adults). Throughout childhood children are accommodating their schemas as they learn to interpret their experiences in ever more complex ways.

Rules such as 'multiplying always makes it bigger' 'division of small by large is impossible' are frequently applied by secondary school pupils. These are correct if limited generalisations from their experience of whole numbers. However, when these are applied to broader domains such as the integers (negative numbers), rational numbers (fractions), or real numbers (decimals), these generalisations are incorrect.

For example, the problem shown in Figure 5.2 was presented to 14/15 year olds (Hart 1981).

*Figure 5.2: CSMS problem probing accommodation of multiplication schema*

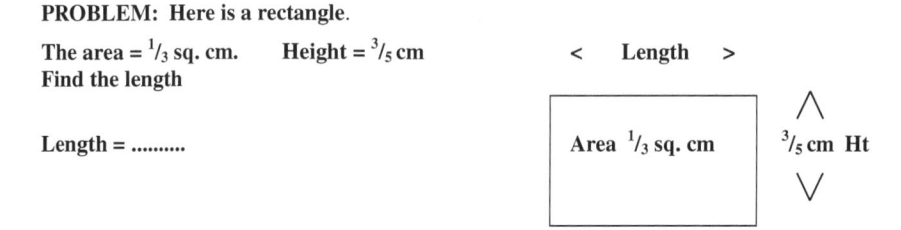

**PROBLEM: Here is a rectangle.**

The area = $1/3$ sq. cm.     Height = $3/5$ cm     < Length >

**Find the length**

Length = ..........

Area $1/3$ sq. cm     $3/5$ cm  Ht

7% of 14 year olds answered correctly, as did 5.6% of 15 year olds. However, the vast majority did not attempt it. Some said: "$3/5$ is bigger than the area, it cannot be done" (Hart 1981: 68 & 76).

This implies that they are still using the multiplication schema based on their experiences of whole numbers. The undeveloped multiplication schema that includes the knowledge 'multiplying makes bigger', must be accommodated when they learn that this is untrue for fractions (smaller than one), or else they will not progress in their mathematical understanding.

There is a simple analogy between learning by means of schemas and the growth of a crab. As a crab ingests food its body weight increases, gradually, by increments. There comes a time when it cannot grow any more, because of its hard shell. The crab grows a larger shell to accommodate its growth, and then discards the smaller shell. As it steps out, it expands and hardens its new shell.

There is a more powerful analogy between the growth of mental knowledge, in the form of schemas, and the growth of scientific knowledge, in the form of theories. According to the widely accepted account of Thomas Kuhn (1970) of the s*tructure of scientific revolutions*, science alternates between two types of development. These are periods of *normal science* and periods of *revolutionary science*. During a period of normal science there is an accepted theory which guides research, and to which new knowledge is added. The analogy with the assimilation of new experience to an existing schema is evident. Thus, for example, the Ptolemaic theory of the planets and sun orbiting the Earth was the basis for astronomy for a thousand years up to the Sixteenth Century.

In this history of astronomy example, Copernicus was disturbed by the increasingly incorrect placing of Easter (up to 11 days late) which resulted from the accumulated errors of Ptolemaic astronomy. This, amongst other things, led to the Copernican revolution in astronomy. And as we all know, it took a long time before Copernicus' theory replaced the old one.

As in this example, a period of revolutionary science is precipitated when a powerful and unsupportable inconsistency between the theory's predictions and fact emerges. After a period of ferment (the revolution) a new theory emerges which explains the previous knowledge in a new way, overcoming the inconsistency. However in the new theory not only are the laws (the relationships between the terms of the theory) different, but the terms themselves are likely to have new meanings. Again, the analogy with the accommodation of schemas is evident. Conflicts between new facts or experiences and a preconception (schema) lead to a cognitive conflict. Cognitive conflicts trigger the accommodation, i.e., adjustment, of schemas. During this process restructuring takes place; concept and their links are readjusted, which changes the learner's concepts and outlook.

This is why it is so difficult to recapture the feeling of ignorance or lack of understanding after a new way of seeing something has been taken on board. We are no longer able to see things the way we did, after our mental schemas have been restructured.

This chapter has looked in a little detail at certain aspects of children's learning of mathematics. In particular, it has explored the theory of learning which has children's construction of their own understanding of mathematics, as its central tenet. This constructivist view of learning does not imply any particular approach to mathematics teaching, other than a general sensitivity to children's existing knowledge, and the provision of opportunities for children to actively extend their understanding (see next chapter). A number of different teaching approaches are compatible with a constructivist view of learning. Indeed Chapter 2 suggested that, to attain all the objectives of the teaching and learning of mathematics, the eclectic approach proposed by Cockcroft (1982) is required. Chapter 6 below discusses in more detail the different varieties of constructivism and their possible implications or emphases for teaching.

*EXAMPLE: MENTAL NUMBER LINE IMAGES*
The following is offered both as an example of a conceptual structure and of the construction of personal knowledge. Around 1980 I picked up a book called The Creatively Gifted at a garage/jumble sale. In it is a diagram of one persons mental

image of the number line. I reproduce it below. When the person whose image it is thought of numbers between 1 and 1000 he brought to mind this picture, and identified the point on the image corresponding to the number. As I read this, I realised with a shock that I too have a mental number-line image. I ran my mind over it from 1 to 1,000,000, and could see it as something like a wire on telegraph poles attached at significant figures, and looping up a hill (the second figure below).

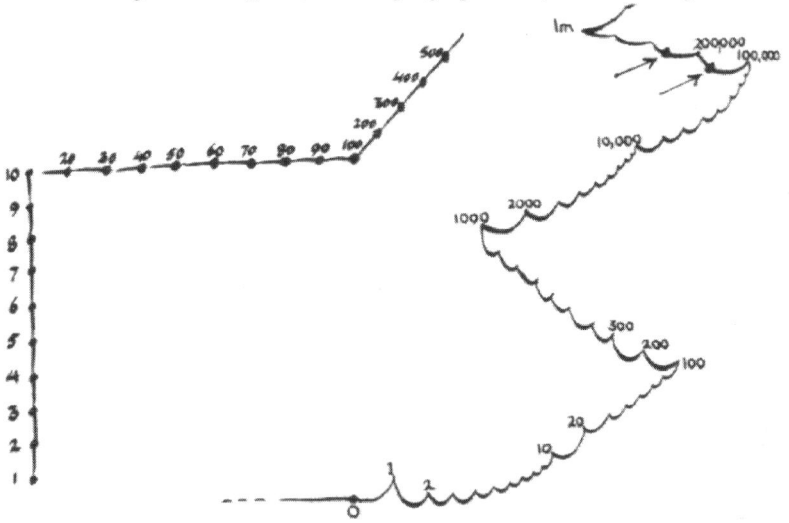

I realised that whenever I think of a number, say 100 000, my understanding of that number included the image of its location on the number line. I also noted that two non-significant (i.e. non-rounded number) points on the line were marked at 239 000 and 186 000. I realised that these represent the distance to the moon in miles and the speed of light in miles per second, and derive from my boyhood fascination with space travel. Thirty years later, the marks were still there.

After doing a little research (Ernest 1983a, 1983b, 1986, 1988a) I found that number line images are not uncommon, and were first studied by the eminent Victorian psychologist Francis Galton (1883) who called them 'number forms'. I reproduce a page of his results below, different images described by the subjects of his study.

What this account of mental number-line images shows is that persons build up their own personal and idiosyncratic knowledge representations, although use in conventional contexts ensures (by a process of correction) that they lead to the same

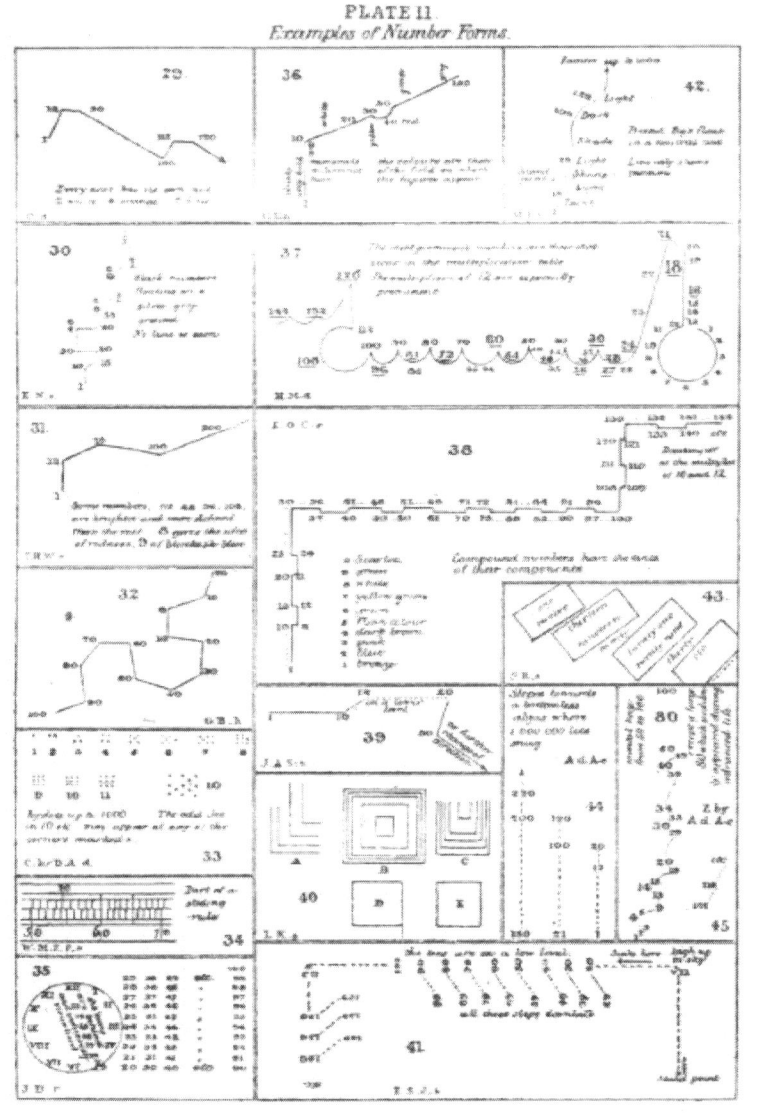

PLATE 11.
*Examples of Number Forms.*

performances. All of the images illustrated above share the key feature of illustrating a linear ordering of the whole numbers, which is their essential characteristic.

Having illustrated and sketched some of the features of the constructivist theory of learning, Table 5.1 contrast some of its key aspects with the passive reception theory of learning, i.e., that knowledge is passively received by the learner.

*Table 5.1: Contrasting theories of learning*

**THEORIES OF LEARNING MATHEMATICS**

| ASPECT | PASSIVE RECEPTION THEORY | CONSTRUCTIVIST THEORY |
| --- | --- | --- |
| LEARNING | Knowledge is passively received and memorised. E.g., learn 9 × table by heart | Knowledge actively constructed by learner from existing knowledge. Learner constructs relationships, e.g., 9×7=63 by association, representation, etc., from existing knowledge and new links |
| INFORMATION | Carries meaning and knowledge | Consist only of symbols to which meaning must be attached by learner |
| ERRORS | Due to faulty memory or carelessness on part of learner | Due to idiosyncratic construction of knowledge - essential for the adjustment of personal knowledge |
| PROBLEM SOLVING | A special application of existing knowledge to a new task | All learning and action is problem solving: Learner must ask herself: What is nature of this task (teachers' request, maths task)? How do I do it? |

According to the constructivist theory of learning, knowledge is actively constructed by the learner from their existing knowledge on the basis of their experiences. This contrasts with the traditional view of learning according to which knowledge is passively received and memorised.

According to the constructivist theory information consists only of symbols and representations to which meaning must be attached by learner. (of course, reading the meaning of others' body language, sounds and speech, as well as pictorial and written symbols, is one of the major learning tasks and intellectual accomplishments of all human beings, and once achieved it seems as if the information carries meaning rather than if learned interpretations are being applied.) In contrast, according to the traditional view information simply carries meaning and knowledge.

According to the constructivist view errors are essential for the correct (i.e., Conventional) adjustment of personal knowledge, and are due to the idiosyncratic (personal) construction of knowledge. In contrast, the traditional view is that errors are due to faulty memory or carelessness on the part of learner.

Finally, the constructivist view regards all learning and action is problem solving. In any task the learner must ask herself: what am I being asked to do? How do I do it? What does the teacher really want? What is the mathematics task? In contrast, the traditional view is that problem solving is a special application of existing knowledge to a new task.

This account shown some of the features and strengths of the constructivist theory of learning. However the account somewhat stereotypes and over-simplifies what it calls the passive reception theory of learning. As I describe below, in recounting Ausubel's (1968) theories, reception learning need not be rote, but can be meaningful learning which leads to the of building up conceptual structures and schemas.

## SCHEMA THEORY AND UNDERSTANDING IN LEARNING MATHEMATICS

In chapter 2 the learning outcomes of mathematics were specified, following Bell *et al.* (1983), to include conceptual structures, which are sets of concepts and their linking relationships - e.g., the concept of place value. Another name for a conceptual structure is a schema. Traditionally schema is the singular term, with schemata the plural, but I follow modern simplified usage of schema (singular) and schemas (plural). In fact schemas are not only relevant to the cognitive domain, but can also store knowledge and records of experiences relevant to the affective and psycho-motor domains, i.e., concerning attitudes, dispositions, bodily movements and perceptual matters. The key thing about conceptual structures and schemas is that they consist of meaningful storing knowledge, information and the memories of experiences.

Earlier I described how facts can become easier to retain in long term memory through mnemonics or links to arbitrary conceptual structures, or through interlinking among themselves (i.e., with other facts) and linking with concepts, skills, etc. Meaningful learning consists in linking new knowledge and skills with existing (or newly created) conceptual structures or schemas.

Ausubel enunciated the principle that a child's (or a person's) pre-existing knowledge, i.e., their schemas, are the most important factor in what the child learns next.

> The most important thing is what child already knows, ascertain that and teach him/her accordingly. (Ausubel 1968)

Secondly, Ausubel clarified what meaningful learning means and invented the term 'meaningful reception' learning to distinguish it from discovery learning, which was the vogue in the 1960s. The following figure (5.1 distinguishes two pairs of opposites: meaningful versus rote learning, and reception versus discovery learning.

Meaningful learning is linked to existing knowledge, it is relational, and schematic. Rote learning is arbitrary (you just have to know it without being able to use reason to reconstruct it), it is verbatim (it must be remembered exactly as learned), it is disconnected (not related to schemas or other existing knowledge of the learner).

*Figure 5.1: Ausubel's distinction between meaningful/rote and reception/discovery*

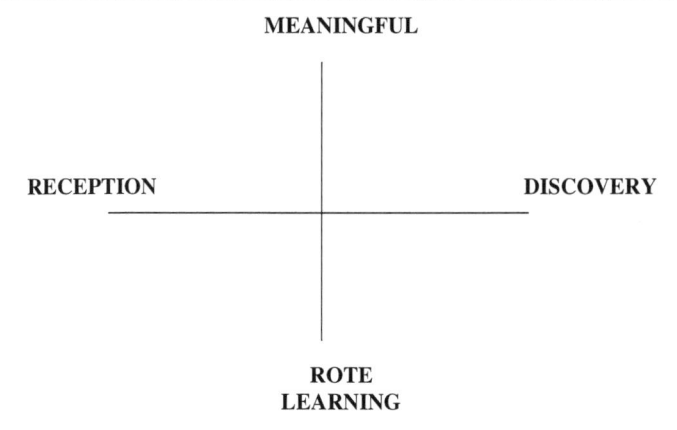

```
                    MEANINGFUL
                        |
                        |
                        |
  RECEPTION             |             DISCOVERY
  _____
                        |
                        |
                        |
                        |
                      ROTE
                    LEARNING
```

Meaningful learning is linked to existing knowledge, it is relational, and schematic. Rote learning is arbitrary (you just have to know it without being able to use reason to reconstruct it), it is verbatim (it must be remembered exactly as learned), it is disconnected (not related to schemas or other existing knowledge of the learner).

Ausubel was keen to show that the opposite of discovery learning is not rote learning, that we can acquire meaningful knowledge by other means than discovering it entirely for ourselves. He defined reception learning as the opposite of discovery learning. It is based on the receiving of knowledge previously formulated by, e.g., the teacher. So when children are shown skills, etc., this can lead to reception learning, if they absorb and remember the procedures involved.

Reception learning can be both meaningful, where the new, fully formulated knowledge is linked by the child to their existing schemas, or it can be rote learning, where the new knowledge must simply be learned by heart without this linking.

Another psychological theorist is Richard Skemp (1976), who made the important distinction between instrumental understanding and relational understanding. Instrumental understanding concerns knowing how to perform a skill, but without being able to explain why it works. Understanding which is purely instrumental may be weak, unmemorable, and non-transferable to new situations, especially if rote-learned. Relational understanding concerns not only knowing how to perform a skill, but also knowing why it works and fitting the knowledge into a larger 'map' or schema.

Skemp (1971) conducted a classic experiment in which two groups of secondary school pupils were given tasks, one characterised as rote learning and the other as schematic or meaningful learning. One group (the control group) was given a series of complex symbols to remember by rote learning, e.g., $\bigcirc \rightarrow \approx$ , and more complex combinations of up to 15 symbols, which had no meaning whatsoever for the pupils. The other group (the experimental group) was given the same symbols to learn, but with explanations for the meanings of individual symbols (e.g., → means to move or travels, ○ means container, ≈ means water), and for more complex combinations of up to 15 symbols (e.g., ○→≈ means a container which travels over water, i.e., a ship; note that these symbols were actually presented in a column, also representing the idea of 'over'). This group were thus encouraged to construct schemas linking the symbols and giving them meanings, i.e., they were given meaningful learning tasks (Skemp 1971: 42).

After two days of instruction both groups were given the same memory test involving these complex symbols. The mean percentage recall immediately after the instruction and subsequently were as follows:

*Table 5.2: Meaningful versus rote learning, results of Skemp's experiment*

| GROUP | Immediate Recall | After 1 day | After 29 days |
|---|---|---|---|
| Experimental (meaningful learning) | 69% | 69% | 58% |
| Control (rote learning) | 38% | 23% | 8% |

Thus immediately after the instruction the experimental group already performed almost twice as well as the control group. After one day over a third of the control had forgotten the material group that they had learned by rote, whereas the experimental group retained their learning. After 29 days all but 8% of the control group had forgotten the material, whereas more than half of the experimental group could still remember and understand the material. This demonstrates that although rote learning can be moderately effective in the short term, meaningful learning is not only better in the short term but over any longer period meaningful learning is far superior in terms of retention. After all, our long term memory mechanisms including the formation and use of schemas and conceptual structures, has evolved expressly for the long term retention of knowledge.

Another theorist James Hiebert (1986) distinguished between conceptual and procedural knowledge in mathematics. Conceptual knowledge is a richly interconnected body of knowledge indicating relationships, i.e., relational or schematic knowledge, as previously discussed above. Procedural knowledge is knowledge of procedures or algorithms. It constitutes knowledge of skills, that is knowing what to do in applying procedures. This partly corresponds to Skemp's idea of instrumental knowledge. However Hiebert's point is not that conceptual knowledge is superior to procedural knowledge in mathematics. Rather it is that conceptual knowledge (or relational understanding) complements procedural knowledge (or instrumental understanding). They are different types of knowledge and we need both in mathematics and throughout our other areas of life and functioning. In this he echoes the philosopher Gilbert Ryle's (1949) distinction between knowing how (e.g., knowing how to use a protractor, or having the skills of teaching) and knowing that (e.g., knowledge of the highway code, or knowledge of education). Know how typically is practical 'how to do it, knowledge, whereas knowing that typically is theoretical or propositional (verbal) knowledge.

Thus, as in Ausubel's distinctions between meaningful/rote learning on the one hand, and reception/discovery learning on the other, we can make a similar distinctions concerning conceptual versus procedural knowledge. Conceptual knowledge by its nature is meaningful, relational and schematic, and is connected to conceptual structures. Procedural knowledge, can, on the one hand, be meaningful, when it is connected to schemas and conceptual structures. It can, on the other hand be rote learned and unconnected to existing schemas, as in the case with Skemp's (1976) instrumental learning. So it is important to maintain the distinction between rote and meaningful learning.

Ultimately, there is widespread agreement that well entrenched knowledge, be it procedural/instrumental or conceptual/ relational, is that knowledge which is well connected to a schema or conceptual structure. Such high levels of embeddedness in conceptual structure is what makes knowledge memorable, meaningful, accessible in a variety of situations, and transferable in application to novel problem situations. (See, e.g., Bell *et al.* 1983, Dickson *et al.* 1984, Resnick and Ford 1981).

Scholars have reached two main conclusions concerning the relationship between conceptual and procedural knowledge. First of all, procedural knowledge can help to develop conceptual knowledge, because "symbols enhance concepts, ... procedures apply concepts to solve problems, ... [and] procedures promote concepts" (Hiebert 1986a: 15-16). Secondly, conceptual knowledge and understanding requires a higher level of cognitive operation than algorithmic or procedural knowledge. The level of cognitive operation is higher both in terms of demands for abstract thought and reasoning (such as defined in Bloom's 1956, Taxonomy) and in terms of the demand on short term memory in terms of the number of working slots, out of the $7 \pm 2$ available, described in Chapter 2 (Slesnick 1982).

**CONCEPT MAPPING**

This and the previous chapter have explored how children can acquire and develop unique personal conceptual structures and schemas for both their conceptual and procedural knowledge. Indeed the main implication of the constructivist theory of learning discussed here and in the next chapter is that all knowledge is organised into unique, idiosyncratic and personal conceptual structures and schemas. An interesting and valuable area of research is into learner's personal knowledge and concept structures. One attempt at this was described above in the Denvir and Brown (1986) research, although in that case the children's conceptual structures were mapped

against a theoretical analysis of the desired connections between knowledge and skills in two digit subtraction.
Theoretical models of cognitive structures and their growth through the twin process of assimilation and accommodation are widespread in the literature. Resnick (1983) offers a developmental model of young children's number understanding which suggests the role of important schemes (e.g., the Part-whole scheme, the Trade scheme) and illustrates them with cognitive maps. Greeno suggests a sequence of steps in the restructuring of learner's schemes for the four arithmetical procedures and illustrates them with conceptual maps (Resnick and Ford 1983: Chapter 8). Resnick and Ford (1983) also suggest models of children's concept maps for area and the transformations of two dimensional shapes.

In the past decade or two there has also been much empirical work in the area of mapping individual's conceptual structures and 'alternative frameworks' or 'alternative conceptions', as they term the field of study in science education (Head 1986). Indeed Pfundt and Duit (1988) have published a bibliography of alternative frameworks and conceptions in science and mathematics education with well over a thousand articles and research reports listed. Many researchers are reporting studies of individual student's conceptual structures in particular mathematical areas, such as in the areas of whole numbers, fractions, geometry concepts, algebraic concepts, and proposing tentative maps of the constituent facts, skills, concepts and their links and interconnections (see, e.g., Erlwanger 1973, Haseman 1991, Novak *et al.* 1983, and Novak 1987, 1990). Such studies usually are based on the clinical interview approach, in which students are presented with a series of carefully planned tasks and questions designed to elicit their ideas, concepts, understandings and their mental representations. Based on the answers derived from the learner, the researchers put together tentative maps of the learner's conceptual structure. These are often shown in two dimensional diagrams illustrating both the knowledge components and their links.

Malone and Dekkers (1984) describe the use of concept mapping as an aid to both teaching and assessing individual's learning. They identify the following feature to be looked for in individual's concept maps: Concept recognition (identifying the relevant concepts in a given topic), Grouping (appropriate linking or joining of concepts), Hierarchy (more central inclusive concepts at top of map, more specific concepts at lower end of map), Branching (the level of differentiation of concepts), and Propositions (reading links as verbs and linked concepts as nouns, the laws or statements specified in the concept map). Malone and Dekkers provide several examples including a concept map for the topic polygons including 13 concepts and their links.

Huerta (1996) similarly describes using concept mapping to elicit, represent and analyse students' conceptual structures for quadrilaterals. He argues that this approach provides information on students' concepts, the relationships between concepts, and links between concepts. The tools he used included a questionnaire on the rectangle, asking pupils to write as many properties as they knew for 1. the sides, 2. the angles, 3. the diagonals, 4. about symmetry and other relations and properties, 5. about the shape (identifying rectangles from a sheet with 17 quadrilaterals of every type in every orientation including 2 stereotypical rectangles, 2 squares and a diamond-orientated square), and 6. trying to give a definition of a rectangle. Huerta also used a questionnaire in which pupils had to link the term rectangle with 10 other named quadrilaterals with one of the relations 'always is', 'sometimes is', 'never is' (where appropriate), and also do the same for the term rhombus. His results enabled him to classify pupils' concept maps according to the meanings they had; whether they used known concepts or exhibited misconceptions; whether their definitions and classification schemes were inclusive (e.g., square is a rectangle) or exclusive (e.g., square is *not* a rectangle).

Leinhardt (1987) studied some primary school aged children (2nd grade) and elicited their ideas and methods for subtraction with 'borrowing' (regrouping). As well as drawing concept maps for the children, she charted their progress over a two term course of instruction and drew multiple concept maps showing the addition of concepts and new links to their cognitive structures (or rather to her hypothesised diagrams representing them) as they developed expertise. She also charted the teacher's concept map for the same topic to compare the understanding of the instructor (expert) and the instructed (novices). Leinhardt and Smith (1985) also charted (through concept mapping) the knowledge of a number of primary school teachers in several mathematical topics.

Clearly these ideas of concept mapping can usefully be applied in researching individual pupils understanding of mathematical facts, skills, procedures and conceptual structures, using concept maps.

> Concept maps find new meanings in the subject and new ways to relate what they already know to the new thing they are learning... [concept mapping] therefore makes learning meaningful (Novak *et al.* 1983: 635)

| *Assignment Suggestion* |
|---|
| Investigate one or more student's understandings and conceptual structures pertaining to a single mathematical topic or subtopic (e.g., the concept of probability; 2 dimensional geometric shapes; concept of a fraction). On the basis of your findings draw a tentative conceptual map of the learner's understanding in the topic. Be sure to justify each element in the map. |
| This topic done in depth with a number of students could easily form the basis for a dissertation study. |

# CHAPTER 6
## THEORETICAL AND PRACTICAL ASPECTS OF CONSTRUCTIVISM

One of the central problems facing mathematics education world-wide, both theoretical and practical, is how to best facilitate the learning of mathematics by students. In the context of this problem constructivism has become one of the dominant theoretical positions in mathematics (and science) education. A growing number of mathematics educators identify themselves as constructivist researchers or teachers. Many positive consequences flow from this, for the position embodies a powerful vision of the active and epistemologically empowered learner, i.e., one who creates and tests their own knowledge. But there are also dangers that follow from an over-enthusiastic or uncritical embracing of constructivism. In this chapter I try to indicate some of the strengths and weaknesses of adopting constructivist views of learning.

### PIAGET'S CONSTRUCTIVISM

It is Jean Piaget's (1953, 1971, 1972) influence which has established constructivism as a leading theory of learning mathematics. His constructivism includes both an epistemology and a research methodology.

Piaget's epistemology (theory of knowledge) has its roots in a biological metaphor, according to which the evolving organism must adapt to its environment in order to survive. The developing human intelligence also undergoes a process of adaptation in order to fit with its circumstances and remain viable. Personal theories or schemas are constructed as constellations of concepts, and are adapted by the twin processes of assimilation and accommodation in order to fit with the human organism's world of experience. Assimilation is the process in which new experiences are interpreted by means of existing schemas. When new experiences cannot be accounted for by an existing schema, the process of restructuring that occurs is termed accommodation, which also seeks to resolve conflicts between existing schemas. Piaget claims that the human intelligence is ordering the very world it experiences in organising its own cognitive structures in these ways.

Piaget's methodology centres on the use of the clinical interview, in which an individual subject is required to perform certain carefully designed tasks in front of, and with prompting and probing from, an interviewer. For example, in a conservation of number task, the interviewer places a line of e.g. 8 counters or chips in front of a child. She then spreads them out to double the length and asks the child if the number of chips is now more, the same, or less. Several sessions are likely to be needed for the researcher to develop and test her model of the subject's understanding. Piaget's clinical interview method is an important contribution to research methodology. It is among the most widely used approaches, especially (but not exclusively) among constructivist researchers.

## Construction

There are a number of different forms of constructivism, and it is important to distinguish them as they vary in terms of theoretical and practical significance. What the various forms of constructivism all share is the metaphor of construction, that is of carpentry, architecture or construction work. This is about the building up of structures from pre-existing pieces, possibly specially shaped for the task. The metaphor describes understanding as the building of mental structures, and the term 'restructuring', often used as a synonym for 'accommodation' or 'conceptual change', contains this metaphor. What the metaphor of construction does *not* mean in constructivism is that understanding is built up from received pieces of knowledge. The process is recursive (Kieren and Pirie 1991), and so the 'building blocks' of understanding are themselves the product of previous acts of construction. Thus the distinction between the structure and content of understanding can only be relative in constructivism. Previously built structures become the content in subsequent constructions.

This indicates the central and shared insight of constructivism: namely that learners, and indeed all persons and higher sensate beings, make sense of their experiences by meaning-making and interpretation. Both the metaphor of construction and this central notion of interpretation and meaning-making are contained in the first principle of constructivism as expressed by its leading living exponent Ernst von Glasersfeld (1989: 182): "knowledge is not passively received but actively built up by the cognising subject". This is the principle of weak constructivism, constituting those positions based on this principle alone. It is also the shared insight of all forms of constructivism; it is just that some also make additional and stronger claims about learning and knowledge.

## Weak Constructivism

Weak constructivism is thus based on the principle that all individual human knowledge is constructed by each individual for herself or himself. This means that all the data that humans process is self constructed, all the way down to the basic level of electro-chemical nerve impulses. Weak constructivism accepts that there is a pre-given world of persons, objects and conventional knowledge, but adopts an agnostic, tentative position about our knowledge of this world. This position is shared with most schools in the modern philosophy of science. For time and again our best theories of the world are shown to be false (Popper 1959). This position enables learner's constructions of meanings to be problematised, without having to raise larger ontological (that branch of metaphysics concerned with existence and essences) and epistemological issues. It does, however, leave the issue of the nature and status of mathematical knowledge unanswered.

Weak constructivism represents a very significant move from naive empiricism in philosophy or the classical behaviourism in psychology of earlier this century. For it recognises that knowing is active, that it is individual and personal, and that it is based on previously constructed knowledge. Just getting student teachers to realise this, by reflecting on 'child methods' in mathematics or alternative conceptions in science, say, represents a significant step forward from the naive transmission view of teaching and passive-reception view of learning many arrive with. A passive-reception view of learning is far from dead, especially amongst politicians and administrators in education. Many government driven curriculum reforms, in Britain at least, assume that the central powers can simply transmit their plans and structures to teachers who will passively absorb and then implement them in 'delivering the curriculum'. Such conceptions and strategies are deeply embedded in the public consciousness, although it may be no accident that they also serve authoritarian powers. (Freire 1972, Ernest 1991b).

An important outcome of this perspective in terms of learning theory (and pedagogy) is that it accounts for student 'error patterns' in mathematics (Ashlock 1982), and idiosyncratic constructions and alternate conceptions in mathematics and science.

**Radical Constructivism**
Although it originates with the work of Piaget (1953, 1971, 1972), and is anticipated by Vico, in its modern form radical constructivism has been most fully worked out in epistemological terms by Ernst von Glasersfeld, in a series of publications over the past 20 years culminating in Glasersfeld (1995). In methodological terms, the leading figure in the area of mathematics education has perhaps been Steffe (1991).

Definitionally, Radical constructivism is based on both the first and second of Glasersfeld's principles. The second is as follows: "the function of cognition is adaptive and serves the organisation of the experiential world, not the discovery of ontological reality." (Glasersfeld 1989: 182). Thus, learners develop their own personal 'world map' which allow them to interpret whatever they experience and also guides their actions in life. This 'world map' is not necessarily a true picture of the world. Indeed Glasersfeld argues that first of all, we could not have a true picture of the world, and secondly, we could not know it even if we had one. For this would require a higher level map encompassing both our understanding and the world it is supposed to represent, and we would have to know that this higher level map is also true, leading us into an infinite regress.

Consequently, "From an explorer who is condemned to seek 'structural properties' of an inaccessible reality, the experiencing organism now turns into a builder of cognitive structures intended to solve such problems as the organism perceives or conceives." (Glasersfeld 1983: 50). Thus what we think is the world in our experiences is actually only our personal model or 'world map'. While our 'world map' must fit with our experiences of reality in life, this no more means that it is an accurate representation of reality than a key, which opens a locked door, is an accurate map of the lock.

According to the Piagetian evolutionary metaphor, the cognising subject is a creature with sensory inputs, furnishing data that is interpreted (or rather constructed) through the lenses of its cognitive structures; it comprises also a collection of those structures all the while being adapted; and a means of acting on the outside world. The cognising subject generates cognitive schemas to guide actions and represent its experiences. These are tested according to how well they 'fit' the world of its experience. Those schemas that 'fit' are tentatively adopted and retained as guides to action. Cognition depends on an underlying feed-back loop. On a larger scale the organism itself and as a whole, is adapting to the world of its experiences, largely through the adaptation of its schemas.

The environment of the radical constructivist knower is real and resists and constrains the knower's actions, but is not known over and above the ways that the knower's schemas fit or fail to fit the world. No match between these schemas and the world is possible; nor could it be verified, if it did exist. Overall, radical constructivism is neutral in its ontology, making no presuppositions about the existence of features of the world behind the subjective realm of experience. The epistemology is consequently wholeheartedly fallibilist, sceptical and anti-objectivist. The fact that there is no ultimate, true knowledge possible about the state of affairs in the world, or about such realms as mathematics, follows from the second principle, which is one of epistemological relativity. As its name implies, the theory of learning is radically constructivist, all knowledge being constructed by the individual on the basis of its cognitive processes in dialogue with its experiential world.

Radical constructivism is a rich theory, that is giving rise to a whole body of fruitful and innovative research (Davis *et al.* 1990, Glasersfeld 1991, Steffe and Gale 1995, Steffe *et al.* 1996, Steffe 1991). Indeed, it is very important in mathematics and science education, where it might be said to represent the dominant epistemological viewpoint. Almost uniquely, it represents an educational paradigm – a research program even – that has been fully developed in that the ontological, epistemological, methodological and pedagogical dimensions have all been extensively treated in the recent literature.

One central criticism which might require a clearer exposition of the relevant aspects of the position, or some revision of radical constructivism is that the account of the cognising subject (the learner) emphasises its individuality, its separateness, and its primarily cognitive representations of its experiences. Its representations of the world and of other human beings are personal and idiosyncratic. Indeed, the construal of other persons is driven by whatever representations best fit the cognising subject's needs and purposes. None of this is refutable. But such a view makes it hard to establish a social basis for interpersonal communication, for shared feelings and concerns, let alone for shared values. Thus the paradigm needs to accommodate these issues by balancing knowing with feeling, and acknowledging that all humans start as part of another being, not separate.

However Glasersfeld has shown in his treatments of aspects of radical constructivism that it is possible to elaborate the position extensively to answer much of this criticism. Each individual's knowledge of other persons, and hence, mediately, the realm of the social, can be consistently construed as constructions of the individual knower. Such an epistemology is self-consistent, and does not fall prey to facile critiques. Similarly, being ontologically neutral, radical constructivism is not solipsistic, as some critics have claimed, putting individuals in a hermetically sealed bubble of their own reality. Nevertheless, it does seem to put up barriers between individuals, and between individuals and the social world. Weak constructivism, by adopting a less stringent epistemology, permits both knowledge and morality to enter from the world. No such ingress is so easily countenanced by radical constructivism.

Radical constructivism is a rich theory which is giving rise to a whole body of fruitful and innovative research into the learning of mathematics (Glasersfeld 1991, 1995, Bauersfeld 1995, Steffe and Gale 1995, Wood, Cobb and Yackel 1995). Authors including these are grappling with the problem of how to accommodate the social in constructivism. Indeed, some of them have been developing a form of social constructivism building on Piagetian or radical constructivist basis for many years (e.g. Bauersfeld 1980, 1994). There is every indication that some of the possible problems of earlier versions of radical constructivism are being overcome, and that overall it constitutes a progressive research programme, in Lakatos' (1978) terms.

## Social Constructivism

Social constructivism is a more recent development in the psychology of mathematics education, and one immediate problem in describing it is that there are several versions of it. The major distinction between these versions is whether they are individualistic or social in orientation; in particular whether they are Piagetian, or based on a socially orientated theory, such as that of Vygotsky (Ernest 1994b). Piagetian versions of social constructivism occur in the recent work of Wood, Cobb and Yackel (1995), Bauersfeld (1995), and others. These forms of social constructivism can be regarded as elaborations of radical constructivism. What can be said is that they do not suffer from the epistemological problems arising from a neglect of the social (Ernest 1994b). It is not yet clear, however, how they accommodate global socio-political problems, addressed e.g. by critical theory.

The second form of social constructivism originated in sociology and philosophy, with inputs also from symbolic interactionism and Soviet psychology. Subsequently it influenced modern developments in social psychology and educational studies, before filtering through to mathematics education (Ernest 1994b). This version of social constructivism is a newer variant of constructivism (Ernest 1991a, 1998a), although in drawing on Vygotskian roots (and the ideas of G. H. Mead) instead of

Piagetian ones, it differs significantly from the other forms discussed above. It is not a form of constructivism *per se* (i.e. based on Piagetian theory) but a form of *social-constructivism*.

The problematique of this form of social constructivism for mathematics education may be characterised as twofold. It comprises, first, an attempt to answer the question: how to account for the nature of public mathematical knowledge as socially constructed? In this, it already goes beyond the exclusive focus on learning of the other forms of constructivism considered above. Second, how to give a social constructivist account of the individual's learning and construction of mathematics? Answers to these questions need to accommodate both the personal reconstruction of knowledge, and personal contributions to 'objective' (i.e. socially accepted) mathematical knowledge. An important issue implicated in the second question is that of the centrality of language to knowing and thought. In the light of this characterisation, social constructivist epistemology has two interconnected parts, concerned with the nature and origin of public and private mathematical knowledge, respectively. Beyond this, social constructivism as an educational philosophy can also be distinguished. I shall return to this below.

The version of social constructivism considered here regards individual subjects and the realm of the social as indissolubly interconnected. Human subjects are formed through their interactions with each other (as well as by their individual processes). Central to this are shared social forms-of-life and located in them, shared language-games (Wittgenstein 1953). The underlying metaphor is that of conversation, comprising persons in meaningful linguistic and extra-linguistic interaction (Harré 1979, Rorty 1979). Thought is understood as internalised conversation, and mind is seen as part of a broader context, the 'social construction of meaning' (Bishop 1985). Mind is viewed as social and conversational because of the following three assumptions. First of all, individual thinking of any complexity originates with and is formed by internalised conversation; second, all subsequent individual thinking is structured and natured by this origin; and third, some mental functioning is collective (e.g. group problem solving). Adopting a Vygotskian perspective means that language and semiotic mediation are accommodated. Through play the basic semiotic fraction of signifier/signified begins to become a powerful factor in the social (and hence personal) construction of meaning, ultimately leading to conscious and voluntarily controlled symbolic thought and action (Vygotsky 1978).

The social constructivist view of experienced reality is that of a socially constructed world which creates (and is constrained by) the shared experience of the underlying physical and social worlds. This socially constructed reality is all the time being modified to fit the constraints of ontological reality, as well as to prestructure perceptions of it according to socially accepted assumptions. Consequently human knowledge can never give a 'true picture' of reality (an insight shared with radical constructivism). The key issue that distinguishes it from the experiential world of radical constructivism is that social reality pre-exists any individual who is socialised into accepting a modified local version as his or her worldview, whilst simultaneously participating in the ongoing development of that social reality, which means that its is never static. In addition, although the singular has been used there are of course multiple social realities corresponding to the different context-bound outlooks of different groups of persons, and most persons inhabit and share several of these social realities, as they engage in multiple forms of life.

The second thrust of the social constructivist epistemology is a fallibilist and social account of the nature of mathematical knowledge as a cultural and historical construct. This is based primarily on the work of Wittgenstein and Lakatos. Wittgenstein (1953) offers the basis of a social theory of meaning, knowledge and mathematics resting on dialogical 'language games' embedded in 'forms of life'; i.e. pre-existing social practices which humans share and within which they act and communicate. (These forms of life include those of the classroom, and those of research mathematics.) In this respect the basis is like Habermas' (1981) theory of communicative action, except that the speech community is not idealised.

Lakatos (1976) offers an important but incompletely developed philosophy of mathematics, which provides a dialectical or conversational theory of mathematics adopted by social constructivism. At the heart of this is his heuristic or Logic of Mathematical Discovery, which is a dialectical theory of the history, methodology and philosophy of mathematics. This is elaborated into a social constructivist dialogical account of mathematics (Ernest 1994a, 1998a)

The social constructivist account interrelates the individual and public aspects of mathematical knowledge production, communication and warranting, and offers parallel accounts in both the educational and research spheres. Within the contexts of mathematics education, individuals use their personal knowledge of mathematics (and mathematics education) to direct and control mathematics learning conversation both (a) to present mathematical knowledge to learners directly or indirectly (i.e. teaching), and (b) to participate in the dialectical process of criticism and warranting of others' mathematical knowledge claims (i.e. assessment of learning).

Similarly, within the contexts of professional research mathematics, individuals use their personal knowledge both (a) to construct mathematical knowledge claims (possibly jointly with others), and (b) to participate in the dialectical process of criticism and warranting of others' mathematical knowledge claims. In each case the individual mathematician's symbolic productions are (or are part of) one of the voices in the warranting conversation.

There is an overall productive/reproductive cycle by means of which mathematical knowledge is created, transmitted and reproduced. Public mathematical knowledge is transmitted in the form of (accepted) texts, which are transformed for educational purposes. Private mathematical knowledge consists in individuals' capacities which are themselves transformed through education. The negotiations through which both public and private mathematical knowledge are transformed are conversational.

Thus conversation is central, not only to the construction of mind, but also to the construction of shared knowledge and the pedagogical or educational philosophy associated with the position. Conversation plays an essential role in the teaching and learning of mathematics, because individual learners develop personal knowledge of language, mathematics and logic through prolonged participation in socially situated conversations of varying types. In the context of mathematics education, teachers structure mathematical conversations on the basis of texts and their own knowledge in order to communicate mathematical knowledge to learners. However, this is necessary, but not sufficient, for such knowledge to be learned. Sustained two-way participation in such conversations is also necessary to generate, test, correct and validate personal mathematical knowledge. The acquisition and use of subjective knowledge of mathematics by individuals are irrevocably interwoven. For only through utterance and performance are the individual construals made public and confronted with alternatives, extensions, corrections or corroboration.

## Implications for Practice

Ultimately, the import of an educational epistemology or philosophy concerns its implications for practice, notably in curriculum and pedagogy. However, there is little in any pedagogy that is either wholly necessitated or wholly ruled out by an epistemology or even an educational philosophy. A pedagogy is based on a set of values, assumptions and an epistemology. But it remains a theory of techniques for achieving the ends of 'communicating' or offering selected knowledge or experiences to learners in a way consistent with these values and assumptions. Thus any so-called 'implications' for practice are not logical deductions, but consistent approaches that seem to 'resonate' or cohere with the background educational philosophy. With this warning, it can be said that the three groups of constructivist perspectives suggest the following valuable pedagogical implications. These include, first of all, the need and value for:

- Sensitivity towards and attentiveness to the learner's previous constructions;
- Diagnostic teaching attempting to remedy learner errors and misconceptions, with cognitive conflict techniques as part of this;
- Attention to meta-cognition and strategic self-regulation by learners;
- The use of multiple representations of mathematical concepts.

Beyond these pedagogical emphases, a number of stronger implications follow from radical and social constructivism, including an emphasis on:

- The subjective construction of meaning: the ability to construct, call up, and enter into the resultant personally imagined 'math-worlds';
- A concern with learner cognitions, beliefs, and conceptions of knowledge;

- A similar concern with teacher knowledge, beliefs, and personal theories and conceptions about subject matter, teaching and learning;
- Knowledge as a whole is problematised, not just the learner's subjective knowledge;
- Methodological approaches are required to be much more circumspect and reflexive: there is no 'royal road' to truth;
- The import of values, purposes and goals underpinning the processes of mathematics education as an intentional activity
- Awareness of the importance of goals for the learner, and the dichotomy between learner and teacher goals;

Social constructivism also picks out the following as features of significance for the mathematics classroom:

- The linguistic basis of mathematical knowledge, and in particular, the role of its special symbolism in mathematics;
- An awareness of the social construction of knowledge; suggesting a pedagogical emphasis on discussion, collaboration, negotiation and shared meanings;
- The social processes involved in the determination, construction and negotiation of mathematical concepts, methods, symbolism, arguments and results;
- The social processes involved in the warranting of knowledge (the assessment of learning);
- Awareness of the importance of social contexts, such as in the difference between folk or street mathematics and school mathematics
- The social and cultural context within which all mathematics occurs, including interpersonal relationships, social institutions and power relations;
- The historico-cultural context of mathematics, the sources and uses of the artefacts, tools and concepts involved.

In this chapter I have explored different versions of constructivism, all of which have helped to reconceptualise the teaching and learning of mathematics. Naturally there is much that I have omitted, including other variants and learning theories. But overall it can be said that, while constructivism may help to reconceptualise teaching, it does not strictly imply or disqualify any teaching approach. Rote learning, drill and practice, and passive listening to lectures can, as they always have, give rise to learning. Active learning can be mental, and so visible inactivity on the part of the learner is irrelevant. Some teaching techniques may possibly be more or less efficient than others, but the constructivist view of learning does not rule out any teaching techniques *in principle*. Nor does it equate to the 'discovery method' or problem-solving teaching approaches (Goldin 1990). Having said this, I have offered some possible pedagogical emphases that fit with each of the three forms of constructivism considered here, even if not strictly implied by them.

# CHAPTER 7

## MATHEMATICAL PROCESSES AND STRATEGIES

The ability to solve problems is at the heart of mathematics. (Cockcroft 1982, par. 249)

Considerable thought has been given to the nature of problem solving in mathematics by a great many authors. Problem solving as an activity in schooling can be traced back to Brownell (1942), Polya (1945) and earlier authors. Gagné (1985) regards problem solving as the highest form of learning, and defined problem solving as "a process by which the learner discovers a combination of previously learned rules ..... which can be applied ..... to achieve a solution for a novel problem situation".

Orton (1992) similarly describes problem solving as a creative process in mathematical thinking.

Problem solving is now normally intended to imply a process by which the learner combines previously learned elements of knowledge, rules, techniques, skills and concepts to provide a solution to a novel situation.

It is now widely accepted that mathematics is not just a predetermined body of knowledge but it can also be a creative activity in which the learner engages (Ernest 1991a, 1998, Davis and Hersh 1980). So it can be claimed that the reason for learning mathematical rules and techniques is to provide the learner with the ability to solve problems creatively. Many also claim that problem-solving is the real essence of mathematics (Orton 1992), as is implied in the quote from Cockcroft (1982) above.

An actual definition of problem-solving is hard to obtain. We do know that problem solving is not a set of routine exercises involving no concern for the mathematical thought processes involved. Polya (1981) gave a definition of problem solving which is precisely concerned with these processes:

To have a problem means: to search consciously for some action appropriate to attain a clearly conceived, but not immediately attainable, aim. To solve a problem means to find such an action.

Charles and Lester (1982) defined a problem as a task where:

1. The person confronting it wants or needs to find a solution.
2. The person has no readily available procedure for finding the solution.
3. The person must make an attempt to find the solution.

Three important points arise from this definition. The student must want to achieve the goal, the student has no way of immediately reaching that goal, the student must make a conscious effort to reach the goal. The idea of a necessary desire to attain the solution was also expressed by Polya (1981: 63).

The desire to solve your problem is a productive desire: it may eventually produce the solution, it certainly produces a change in your mental behaviour.

Thus it can be seen that solving problems is one of the most important activities in mathematics, representing the creative use of mathematical knowledge in novel or non-routine applications. The general strategies distinguished by Bell *et al.* (1983), and briefly discussed in chapter 2 above, play a key role in problem solving. General strategies are methods or procedures that guide the choice of which skills or knowledge to use at each stage in problem solving.

Problems and problem solving are central both to the history of mathematics and the psychology of mathematics education. In both areas they can be seen to stimulate knowledge growth. Historical problems, such as the Königsberg Bridge Problem, which stimulated Euler to create Topology (Wolff 1963) is also used to introduce students to network theory in today's classroom. However the most striking parallel is in the realm of mathematical processes and strategies. Pappus distinguished between analytic and synthetic problem solving methods. Almost two thousand years later the distinction was used by psychologists to distinguish different levels of cognitive processing (Bloom 1956; see chapter 2 above). More generally, methodologists of mathematics such as Descartes and Polya (1945) have offered systems of heuristics (helpful problem solving strategies) which have driven research in problem solving in psychology and mathematics education (Groner *et al.* 1983). Problem solving research in mathematics education still has much to learn from the history of mathematics, and in the era of computers that history continues to grow and evolve new methods.

## POLYA'S PROBLEM SOLVING HEURISTICS

The most influential account of general problem solving strategies or heuristics in modern times is that of Polya (1945). He defines heuristic as the arts of invention or 'serving to discover'. Heuristics is thus the art of discovery, consisting of methods for finding out, including guided trial and error, and depending on past experience. Heuristics are those general strategies which while they are often helpful in solving problems, by the nature of the undertaking they are unpredictable and offer no guarantee of success. Thus heuristics are unlike procedures or algorithms which only have to be applied correctly to give the correct and anticipated outcome.

George Polya was a creative mathematician who devoted much of his mature career exploring the methods which mathematicians use in solving mathematical problems. He published a number of books on the subject including Polya (1954, 1981) in which he suggests means of teaching problem solving strategies to secondary school and university students. However it is his first book (Polya 1945) that is the best known, and which has been the most influential in mathematics education and psychological research. In this book he distinguishes four stages of problem solving, as follows.

*Table 7.1: Polya's (1945) four stages of problem solving*

| STAGE | PROCEDURE |
|---|---|
| *1.* | *UNDERSTANDING THE PROBLEM* |
| *2.* | *DEVISING A PLAN* |
| *3.* | *CARRYING OUT THE PLAN* |
| *4.* | *LOOKING BACK* |

He suggests that this problem solving approach involves four stages: 1) understanding the problem, 2) devising a plan, 3) carrying out the plan and 4) looking back.

Polya's first stage is that of understanding the problem. This involves making sense of the problem itself, to represent it in a more comprehensible form, perhaps. This first stage may appear to be simple or trivial, however it includes activities such as drawing a diagram to represent the problem or translating it into a mathematical sentence. It also includes decisions about whether or not the question provides sufficient information or any irrelevant information. Any fine-grained analysis of problem solving must begin with the comprehension of the problem situation.

The second stage involves deciding how to approach the problem, what general strategies to apply, and indeed either explicitly or implicitly planning how to approach the problem. The third stage is to apply the strategies in an attempt to solve the problem. These two middle stages are when creativity, insight and inventiveness may be most required by the student.

The fourth stage is to review the solution obtained, to check or validate it. It does not just involve checking the end result, but also includes considerations about whether the problem should be extended or the solution generalised, and to consider implications of the problem. The whole process is cyclic, and the problem solver can be expected to repeat the cycle several, or even many times (incompletely in some cycles), possibly revisiting all of the preceding stages from each stage.

The problem solving process in mathematics can be shown by means of a spatial analogy, as a journey from a starting point to a desired but unknown destination. Figure 7.1 illustrates this. It shows understanding the problem as getting 'inside' the problem; how the first steps towards solution involve overcoming the difficulties surrounding the problem, and that several attempts might e needed. Next it shows how attempts to solve must overcome barriers to the solution. Finally it illustrates how there can be more than one route to the solution.

*Figure 7.1: The problem solving process as a journey*

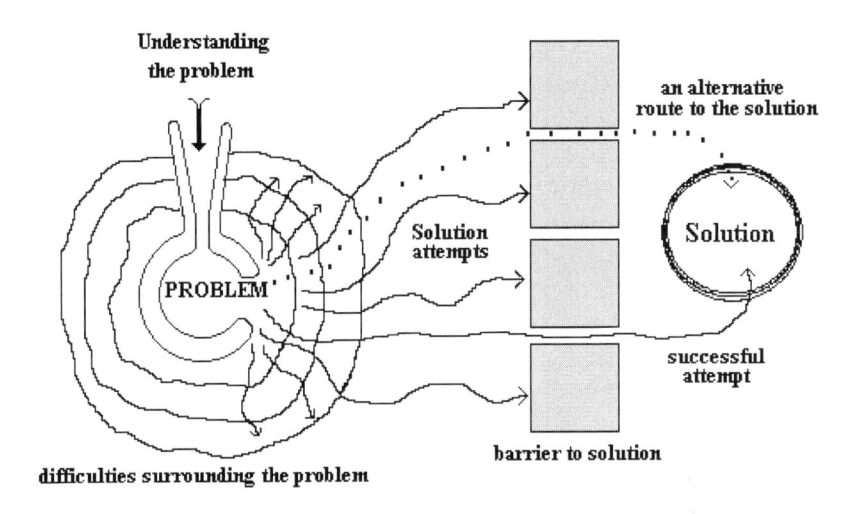

The key feature of problem solving (and investigative work) in mathematics is that it is the *process* of doing mathematics, and not the answer or *product* which is the ultimate educational focus (although the answer or *product* may be the mathematical focus). The process of doing mathematics concerns the development of *mathematical thinking*. Of course, mathematical thinking requires facts, skills and concepts to be known, and it can also involve the use of apparatus, diagrams and pictures. However, the key feature of mathematical thinking is that it is creative enquiring thought, involving strategies and processes such as: imaging (creating mental images), representing, symbolising, explaining, describing, discussing, hypothesising, generalising, taking special cases, classifying, interpreting, rule-making, and proving.

Before exploring the literature on problem solving and mathematical processes it is as well to clarify what is meant by the term 'problem' in mathematics. Webster's 1979 dictionary offers two definitions or of the term which indicate a divergence of meaning.

- *Problem definition 1:* In mathematics, anything required to be done, or requiring the doing of something.
- *Problem definition 2*: A question that is perplexing or difficult

Definition 1 thus concerns any set task, or activity. Typically (and certainly historically) this is often a routine task or exercise. In contrast, definition 2 concerns non-routine tasks which are perplexing or difficult, and which require creativity or the novel use of known facts, skills and procedures for their solution.

LeBlanc (1982) similarly suggests two different types of problem. Firstly he describes the "textbook problem" which the student solves using various previously learned algorithms or formulas. Thus the "textbook problem" is merely a reinforcement of knowledge or skills acquired earlier, and corresponds to definition 1 above.

Secondly he distinguishes the "process problem" which is solved by using certain strategies decided upon by the student, thus requiring more thought and active participation on the part of the student. This corresponds to, but is a more restricted case, of the second definition given above.

This distinction cannot be drawn in absolute or learner independent terms. For what one learner finds perplexing, another may find routine. It is also the case that what one learner finds perplexing at a given time, she will often find routine later. Thus finding how many trips are needed to ferry 20 cars across a river on a ferry that can take just 4 cars may be a non-routine problem during the early years, but soon becomes a routine 'word problem' for most primary and secondary school age children. The important issue here is that for problems of the second type, where no routine procedure is to hand, general strategies are called for to suggest possible ways of proceeding: i.e. which processes and skills to employ.

Of course perplexing and non-routine mathematical enquiry and problem solving tasks encompass a wide range of activities, including pure mathematics, applied mathematics and multidisciplinary project work, such as:

- solving internally oriented mathematical problems,
- solving 'real' mathematical problems relevant to some aspect of the learner's experience outside of the mathematics classroom or home environment,
- open-ended mathematical investigations beginning with mathematical starting points,
- the investigation of practical mathematical situations by modelling or with the aid of apparatus,
- the carrying out of project work on some interdisciplinary or multidisciplinary theme, such as an aspect of the environmental, of which mathematics is just one part.

So the issue of problem context as well as that of cognitive demand and the general strategies and processes involved are important.

Although distinctions between problem types depend on various factors, Lester and Charles (1982) offer a useful classification of problems into six different types, They distinguish: drill exercises; simple translation problems; complex translation problems; process problems; applied problems; puzzle problems. They illustrate these six types in the following way:

## 1. Drill Exercise
$$346 \\ \underline{\times 28}$$

The first of these problems, the drill exercise, is the simple application of a known algorithm. This is a simple routine task and relates to LeBlanc's "textbook problem" discussed previously, in which a known algorithm is applied and a previously learned skill is reinforced.

## 2. Simple Translation Problem
*Jenny has 7 tropical fish in her aquarium. Tommy has 4 tropical fish in his aquarium. How many more fish does Jenny have than Tommy?*

## 3. Complex Problem
*Ping-Pong balls come in packets of 3. A carton holds 24 packs. Mr. Collins, the owner of the sporting goods store, ordered 1800 Ping-Pong balls. How many cartons did Mr. Collins order?*

The second and third 'translation' problems require the student to translate the problem into a mathematical expression, either mentally or in written form. With the simple problem only requiring one of these translations and the complex problem two or more. Again, these two types of problem could relate to the 'textbook' problem, in that the student is applying a learned technique, reinforcing the understanding of a learned mathematical concept.

## 4. Process Problem
*A chess club held a tournament for its 15 members. If every member played one game against each other member, how many games were played?*

The fourth type of problem lends itself to simplification, the student could solve the problem for 2 members, 3 members, 4 members, etc., and then identify patterns in order to solve the original problem.

## 5. Applied Problem
*How much paper of all kinds does your school use in a month?*

The applied problem relates to a realistic situation and the solution requires the student to use all of their acquired skills concepts and procedures. Thus they must decide on a plan of action in order to obtain a solution to the problem, making this type of problem a "process problem".

## 6. Puzzle Problem
*Draw 4 straight line segments to pass through all 9 dots in Figure 1. Each segment must be connected to an endpoint of at least one other line segment.*

Lastly we have the puzzle problem, which can usually only be solved by a lucky guess or by thinking of an alternative, slightly unusual way in which to approach the problem. These problems are different to all of the others, a methodical or analytical approach to the problem is often of little help when looking for a solution. More often than not, the solution to the problem involves some sort of trick which some students may find frustrating. In this case the student may not wish to find the solution, which was one of the conditions of the definition of a problem. Furthermore, succeeding in a puzzle problem will not normally result in the development of transferable problem solving strategies. More commonly it merely involves learning a trick.

## General Problem Solving Strategies or Heuristics

Publications on problem solving strategies go back to the Greek mathematician Pappus who distinguished analytic and synthetic methods in geometry. More recently, philosophers have attempted to systematise scientific creation in ways that are forerunners of mathematical heuristics. Bacon (1960) proposed a method of induction for arriving at hypotheses, which were then to be subjected to testing. In order to facilitate the genesis of inductive hypotheses, he proposes the construction of systematic tables of results or facts, organised to show similarities and differences. Such proposals, published in 1620, anticipate the heuristics of modern researchers on mathematical problem solving, such as Kantowski, who specified "Heuristic processes related to planning... Searches for pattern... Sets up table or matrix" (Bell et al. 1983: 208).

In 1628 Descartes (1931) published a work embodying 21 'Rules for the direction of the mind'. This proposes further heuristics, many explicitly directed at mathematical invention. These include

- simplification of questions
- sequential enumeration of examples to facilitate inductive generalisation
- the use of diagrams to aid understanding
- symbolisation of relationships
- representation of relationships by algebraic equations
- simplification of equations

These anticipate many of the heuristics published 350 years later as aids to teaching problem solving, such as in Polya (1945), Mason et al. (1982) and Burton (1984).

The Cockcroft report (1982), drawing on the research of Bell et al, distinguished general strategies as one of the vital learning outcomes of school mathematics. General strategies, coupled with other components of mathematical knowledge including facts, skills, and concepts, are what is necessary to solve mathematical problems, and this has always been recognised as a central feature of successful mathematics learning.

Currently, in the British (i.e. English and Welsh) National Curriculum in mathematics, general problem solving strategies are an explicit objective in the assessment of mathematics teaching and learning. The attainment target Ma1 *Using and applying mathematics* is made up of three strands intended to represent the use of mathematical processes, namely, applications, mathematical communication, and reasoning, logic and proof. Despite its name, the last of these is more concerned with strategies like generalisation than with proof.

Overall, general problem solving strategies within mathematics consist of high level procedures which guide the choice of what skills or knowledge to use at each stage in general mathematical problem solving.

Some of the general strategies of this type identified in HMI (1985) are the following:

- Trial and error methods
- Simplifying difficult tasks
- Looking for pattern
- Reasoning
- Making and testing hypotheses
- Proving and disproving

General strategies are known under a variety of names, such as problem solving strategies, mathematical processes, processes of inquiry in mathematics, investigative strategies, problem solving heuristics, etc. A key feature of general strategies is that they are potentially fruitful approaches that often work but are not guaranteed to do so. By its nature, problem solving, involving attempted solutions of non-routine problems, can offer no guarantees. As the history of mathematics illustrates, while some problems are rapidly solved, some remain unsolved for centuries before being solved (e.g. The Four Colour problem, and Fermat's Conjecture, which have both of been solved in the past 25 years).

My observation of students engaged in solving mathematical problems has revealed the use of the following general strategies.

Understanding the problem (by discussion)
Re-reading the problem
Drawing a diagram
Representation of problem by symbolisation
Generating examples
Obtaining data
Tabulating the data
Make table of results
Put the results in table in a suggestive order
Searching for a pattern among the data
Conjecture a relationship from table
Making conjectures
Test a conjecture
Thinking up different approaches

Trying them out
Simplifying the problem
Try to solve a simpler problem
Hold one variable constant (at a small fixed value) and exploring the values of the other variable
Trying other approaches
Justifying answers

**Levels of problem solving strategies**

Three levels of general strategies can be distinguished:

1. general strategies applicable across the curriculum or beyond mathematics
2. general problem solving strategies within mathematics
3. domain or topic specific strategies in mathematics

These correspond to three different types of higher order cognitive skills distinguished by Champagne (1992): 1. generic cognitive skills, 2. discipline specific cognitive skills, and 3. Task-specific cognitive skills.

**3. Domain or topic specific strategies in mathematics**

This category concerns domain or topic specific strategies in mathematics, the most concrete or specific area of strategies used in mathematical problem solving. These are strategies which determine (or help to guide) which domain or topic specific facts, skills, knowledge and procedures are brought to bear in solving problems in a topic such as algebra, trigonometry, calculus, geometry, probability, statistics, or mathematical computing. Some illustrations of topic specific mathematical strategies are shown in table 7.2.

*Table 7.2: Topic specific strategies in mathematics*

| | |
|---|---|
| Algebra | How to solve a linear equation |
| | How to find the point of intersection of quadratic & linear graph |
| | How to solve a quadratic equation |
| Trigonometry | How to solve trigonometric equations |
| Calculus | How to integrate a function $f(x)$ with respect to $x$ |
| | How to differentiate a function $f(x)$ with respect to $x$ |
| Geometry | How to demonstrate that two triangles are congruent |
| | How to find out what 2-D geometric transformation of the plane a 2×2 matrix represents |
| Probability | How to find out the number of possible outcomes in a game |
| Statistics | How to determine if data collection result statistically significant |
| Computing | How to access data base |
| | How to debug a Logo program |

Most such strategies are non-transferable within mathematics itself, since they relate to a particular topic area and certain class of problems. Such strategies are normally acquired whilst learning and solving problems within the given topic area. Such strategies are not much discussed in the problem solving literature, but acquiring proficiency in them is an important part in learning any given topic.

Domain specific strategies are the lowest level of general strategies. The next level down in terms of generality takes us out of the area of strategic thinking, for it is made up of topic specific mathematical procedures, algorithms and skills. These guarantee an output when given appropriate input, and thus are not heuristic problem solving strategies.

## 1. General strategies across the curriculum

Problem solving strategies and general methods of inquiry are employed in many areas of study and practice. Within the school curriculum some of the processes of enquiry and general strategies used in other subject areas resemble those of mathematics. For example, investigational strategies and processes are widely used in school science, and problem solving strategies are a central part of Design and Technology. So in mathematics we must not forget that learners acquire general strategies in many classroom contexts beyond that of mathematics.

Similarly, many academic areas, such as educational research involve general researcher strategies analogous to those used in mathematical problem solving (see, e.g., Ernest 1994). Some of the processes that teachers and educationalists expect to see employed in school-based problem solving projects and investigations are precisely those that are needed in the context of research. These include: formulating problems, searching the literature, collecting and tabulating data, making conjectures and checking them, describing particular cases as accurately as possible, etc. Thus carrying out and writing up an educational enquiry is very much like solving a problem or conducting an investigation in mathematics, science, technology, or in other subject areas.

It is remarked in a number of areas, including mathematics, that the process of problem solving is not linear but has the form of a problem solving cycle. This can be illustrated as follows (Figure 7.2).

*Figure 7.2: The problem solving cycle*

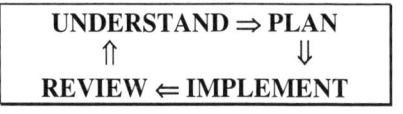

| UNDERSTAND $\Rightarrow$ PLAN |
| :--- |
| $\Uparrow$ $\qquad\qquad\qquad$ $\Downarrow$ |
| REVIEW $\Leftarrow$ IMPLEMENT |

This illustrates how deliberate problem solving can be seen as a sequence of stages. First there is the understanding of the problem. This is followed by conscious planning how to solve it. Then the plan is implemented. Finally, the outcome is evaluated and reviewed. However the process is not over yet, because a first attempt at solution will usually need to be reworked and improved, and the cycle continued a few more times. Each stage can lead to those that follow in the cycle. Planning, implementing or reviewing may lead to a renewed need to better understand the problem. Implementing a plan, reviewing the outcome, or better understanding can lead to the need to better planning how to solve the problem, and so on.

There is also a striking analogy between the problem solving cycle and computer programming which has been remarked by Noss (1983) and Ernest (1988b), as shown here in Table 7.3.

*Table 7.3: The analogy between problem solving cycle and computer programming*

| PROBLEM SOLVING | PROGRAMMING |
| --- | --- |
| **Understand problem** | **Analyse problem** |
| **Devising a plan** (analytic or synthetic strategy) | **Write program** ('top-down' or 'bottom-up' approach) |
| **Execute plan** | **Run program** |
| **Check solution and review plan** | **Debug and modify program** |

Each stage of Polya's problem solving process has a direct parallel with a comparable stage in constructing a computer program. This is shown in the table. In addition, problem solving can be either analytic or synthetic, which provides an analogy with computer programming strategies. An analytic problem solving approach is based on a single overall master plan or strategy, such as working backwards from the solution. A synthetic approach is more tactical, combining a series of little steps until they reach the desired solution. Analytic and synthetic problem solving strategies have a strong analogy with 'top-down' and 'bottom-up' computer programming strategies, respectively. A 'top-down' approach in computer programming works down from a general master plan like the analytic approach in problem solving. The 'bottom-up' approach combines smaller computer procedures, like the synthetic approach in problem solving.

I should stress that this analogy is between mathematical problem solving and *writing* a computer programme. This is a creative and thoughtful activity, unlike *running* a computer programme, which is a purely mechanical process.

A consideration of general (i.e., cross-curricular) strategies raises the issue of the transfer of learning. A growing body of research suggests that not only what students learn, but where they learn it that is important, and that the transfer of learning between contexts often does not take place. As long ago as 1912 Thorndike showed that learning the traditional Euclidean Geometry and its proof strategies did not necessarily transfer to improved reasoning skills - either logical or problem solving (see Kilpatrick 1992).

One theoretical approach to the problem of context and transfer is that of situated cognition. This has been developed by J. S. Brown, J. Lave, A. Collins, L. Resnick, B. Rogoff, and others in the USA. They claim that what we learn is only learned relative to the social context of acquisition. That is, the social context, or 'frame', through its associations for the learner, which triggers or elicits the skills and knowledge from long term memory. These writers claim that knowledge is stored separately according to the context of learning. A consequence is that transfer will not occur unless links or bridges between different contexts and domains of mathematical knowledge are constructed or made by some creative act of mentally linking what was learnt separately. (See, e.g., Lave 1988).

Partial confirmation of this approach can be found by reflecting on the divisions between subjects in secondary school. Even though they are studied in the same school (although often in separate classes), students usually do not connect what they learn in mathematics, science or technology, let alone with that from non-scientific subjects, even if the same mathematical knowledge and skills (from our expert perspective) are involved.

## 2. General problem solving strategies within mathematics

The third class of general strategies that can be distinguished, which is intermediate in level between topic specific and cross-curricular strategies is that of general problem solving strategies within mathematics.

As discussed above, the classic work on problem solving in mathematics is Polya's (1945) *How to Solve it*, and this is primarily directed at this level of general strategies. A number of other authors have offered a similar analysis of problem solving. For example, Leone Burton's (1984) 'Thinking things through' offers a strong parallel with Polya's heuristics (Table 7.4).

*Table 7.4: Comparison of Polya and Burton's heuristics*

| L. Burton | G. Polya |
| --- | --- |
| (Thinking things through) | (How to solve it) |
| Entry | Understanding the problem |
| Attack | Formulating a plan |
|  | Carrying out the plan |
| Review / Extension | Looking back |

The only major difference is that Burton combines the stages of 'formulating a plan' and 'carrying out the plan' in her single phase of 'Attack'. Based on her three part analysis, Burton offers a detailed set of heuristic or general strategies as follows (Table 7.5).

*Table 7.5: Set of General Strategies / Heuristics from Burton (1984)*

**Entry**
1.1 Explore the problem
1.2 Make and test guesses
1.3 Define terms and relationships
1.4 Extract information
1.5 Organise the information
1.6 Introduce a representation
1.7 Introduce a form of recording

**Attack**
2.0 Be systematic
2.1 Search for relationships
2.2 Analyse relationships
2.3 Make simplifying assumptions
2.4 Find properties the answer will have
2.5 Try particular cases
2.6 Adjust guesses
2.7 Formulate and test hypotheses
2.8 Try related problems
2.9 Control variables systematically
2.10 Use one solution to find others
2.11 Work backwards
2.12 Focus on one aspect of the problem
2.13 Eliminate paths
2.14 Partition the problem into cases
2.15 Reformulate the problem
2.16 Upset set
2.17 Develop the recording system
2.18 Change the representation
2.19 Make a generalisation

**Review- Extension**
3.1 Check
3.2 Look back
3.3 Communicate
3.4 Find isomorphic problems
3.5 Extend to a class of problems
3.6 Create different problems

Perhaps the most novel heuristic, going beyond those that have figured above, is 2.16: Upset set. 'Set' is the conjunction of often unexamined or taken-for-granted learner-assumptions concerning a situation or problem. Sometimes to solve a problem, just as sometimes in the history of mathematics, it is necessary to confront and challenge these taken-for-granted assumptions. For example, the following puzzle illustrates this.

*Figure 7.3: Sample puzzle*

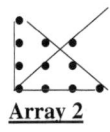

**Array 1**    **Array 2**

The puzzle consists of drawing a continuous line made up of at most four straight line segments through all 9 points of Array 1. Young learners often make many attempts to join them. The 'set' that needs to be upset is the assumption that the lines will be interior to the 3×3 array indicated in the problem. In fact, the solution involves using points beyond this grid, and uses all the points in Array 2, comprising 2 'edges' and 2 'diagonals' making up an arrow head pointing SW.

## METACOGNITION

The central point I have been making is how important the strategies of problem solving are in mathematical thinking. But this analysis can be taken further. The main intellectual activities involved in problem solving can be divided into the cognitive and metacognitive. Cognitive activities include using and applying facts, skills, concepts and all forms of mathematical knowledge. They also include applying general and topic specific mathematical strategies, and carrying out problem solving plans. These are the kind of things that I have been discussing above. Metacognitive activities however involve planning, monitoring progress, making effort calculations (e.g. "Is this approach too hard or too slow?"), decision making, checking work, choosing strategies, and so on. Metacognition (literally: above or beyond 'cognition') is about the management of thinking.

According to Flavell (1976: 232)

> 'Metacognition' refers to one's knowledge concerning one's own cognitive processes and products and anything related to them. ... metacognition refers, among other things, to the active monitoring and consequent regulation and orchestration of these processes in relation to the cognitive objects or data on which they bear, usually in the service of some Concrete goal or objective.

The following Table 7.6 shows some of the mental activities involved in problem solving.

*Table 7.6: Mental activities in problem solving*

| META COGNITIVE ACTIVITIES |
|---|
| • Planning |
| • Monitoring progress |
| • Decision making |
| • Checking |
| • Choosing strategies etc. |
| COGNITIVE ACTIVITIES |
| • Carrying out plan |
| • Applying strategies |
| • Using skills, knowledge etc. |

The straight forward cognitive activities include carrying out or implementing a plan, applying general strategies, using skills, knowledge, and so on. However the meta cognitive activities include planning, controlling and monitoring progress, decision making, choosing strategies, checking answers and outcomes, and so on.

Fernandez *et al.* (1993) suggest that Polya's model of problem solving can be amended to show the role of metacognition, as follows:

*Figure 7.4: Polya's problem solving model amended to include metacognition*

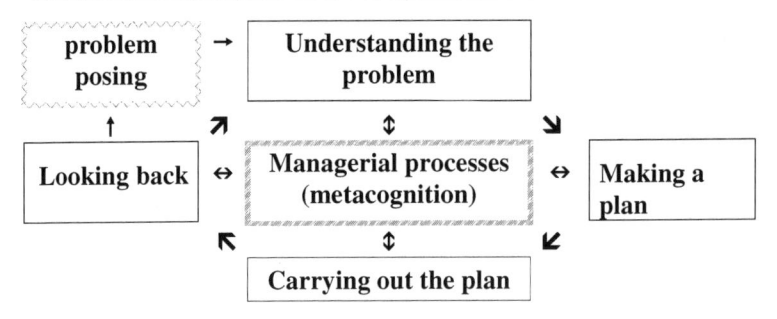

```
   problem      →    Understanding the
   posing              problem
      ↑         ↗          ↕          ↘
   Looking back  ↔   Managerial processes   ↔   Making a
                      (metacognition)             plan
                ↖          ↕          ↙
                    Carrying out the plan
```

This shows, first of all, Polya's problem solving cycle in four stages: Understanding the problem, Making a plan, Carrying out the plan, Looking back, both starting and finishing with an additional activity problem posing (see the discussion of investigation in mathematics below). Secondly, it shows how each stage is subject to managerial (metacognitive) processes, which direct and control problem solving activity.

To help raise thinking level to that of meta-cognition questions can be used. Meta-cognitive questions focus the attention of the problem solver on reflecting on and controlling progress towards the problem goal for which to aim and partly by aiding the recall of associated general strategy. (The organising questions shown in Table 7.7 are taken from Burton 1984).

*Table 7.7: Burton's organising questions*

| | |
|---|---|
| *Entry* | 1.1 What does the problem tell me? |
| | 1.2 What does the problem ask me? |
| | 1.3. What can I introduce to help get started? |
| *Attack* | 2.1 Can I make connections? |
| | 2.2 Is there a result which will help? |
| | 2.3 Is there a pattern? |
| | 2.4 Can I discover how or why? |
| | 2.5 Can I break down the problem? |
| | 2.6 Can I change my view of the problem? |
| *Review - Extension* | 3.1 Is the resolution acceptable? |
| | 3.2 What can I learn from the resolution? |
| | 3.3 Can I extend the resolution? |

Polya also suggests a number of general strategic questions put so as to help guide the student to help develop their meta-cognitive skills.

*Table 7.8: Polya's meta-cognitive questions*

| PROCEDURE | Meta-cognitive Questions |
|---|---|
| UNDERSTANDING THE PROBLEM | Determine: What is the unknown? What are the data? What is the condition? |
| DEVISING A PLAN | Determine: Do you know a related problem? Could you restate the problem? Can you solve a more accessible similar problem? A more general problem? A part of the problem? Can you vary the conditions? Did you use all of the data? |
| CARRYING OUT THE PLAN | Can you check each step? Can you prove each is correct? |
| LOOKING BACK | Can you check the result? Etc.? |

To a large extent the questions indicated here are metacognitive: by asking these the aim is to encourage the student solver to take more control over the way she is attempting to solve a problem. Less experienced solvers are often impulsive - launching themselves down an avenue of attack with little reflection or evaluation of the progress that is being made towards the overall goal of solving the problem.

To illustrate the importance of metacognition in problem solving and the psychology of learning mathematics I will draw on the results of an important programme of research. In reporting his research, Schoenfeld (1992) contrasts the typical pattern of work of novice and expert problem solvers. To do this he analysed problem solving activity into six levels or stages of work, building on Polya's categories. These are: Read, Analyse, Explore, Plan, Implement, Verify. These can be shown (Table 7.9) as corresponding to Polya's stages, but include an additional impulsive (unplanned) stage of exploring the problem.

*Table 7.9: Schoenfeld's analysis of problem activity contrasted with Polya's stages.*

| UNDERSTANDING THE PROBLEM | Read, Analyse |
|---|---|
| SEEKING TO SOLVE WITHOUT PLAN | Explore |
| DEVISING A PLAN | Plan |
| CARRYING OUT THE PLAN | Implement |
| LOOKING BACK | Verify |

This new category is exploration, meaning seeking to solve without plan. This is something that is neither planned nor recommended (at least not to excess) but is something that is frequently observed. Exploration can be valuable for enriching understanding and providing information on which to base planning. But when this stage of activity is persisted in, typically by novice solvers lacking self-regulation or metacognitive skills, it usually leads to failure.

The following diagrams illustrate this. Figure 7.5 shows the time graph of an inexperienced solver's attempt to solve a non-routine problem. After one minute of reading the question, there is an unbroken 19 minutes of unreflective exploration of the problem. The experienced mathematician problem solver however, in attempting to solve a problem, works at all six levels or stages (Figure 7.6), and goes through the problem solving cycle twice. Furthermore, she asks herself questions out loud at the points indicated with little inverted triangles.

Figure 7.7 shows the typical result of an explicit program of training students in monitoring and controlling their problem solving processes. Schoenfeld does this by intervening during problem solving, and asking students to answer: "What are you doing?" (describe it precisely), "Why?" (How does it fit into the solution?) "How does it help?" (What will be done with the outcome?). Schoenfeld's experiments and results were with a sample of university mathematics students. Similar outcomes were achieved by Lester working with 11 years olds (also reported by Schoenfeld 1992). However the development of self-regulatory or metacognitive skills in complex subject-matter domains is difficult, and often involves overcoming or unlearning learned behaviour patterns.

What has been shown is that an explicit program of teaching, over a term say, which concentrates on the development of metacognitive skills during practical problem solving, can result in expert-like behaviour. Others have similarly reported improved performance in mathematical problem solving based on cognitive monitoring and strategies to encourage metacognitive activities (e.g., Garofalo and Lester 1985).

*Figure 7.5: Time graph of a typical student attempting to solve a non-routine problem.*

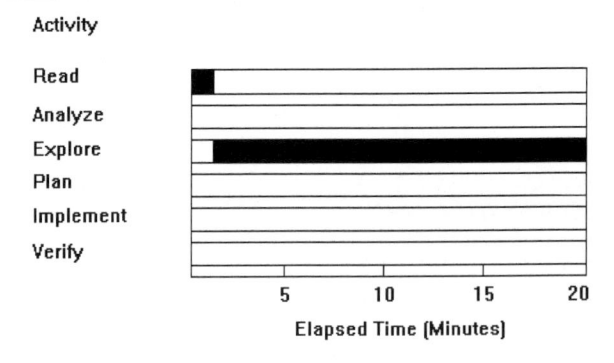

*Figure 7.6: Time graph of a mathematician working out a difficult problem*

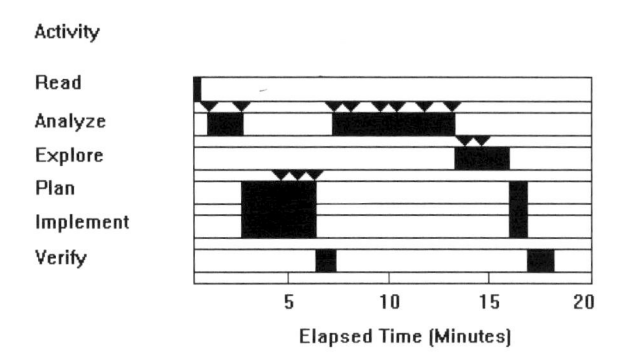

*Figure 7.7: Time graph of a solution attempt after explicit training in monitoring and control.*

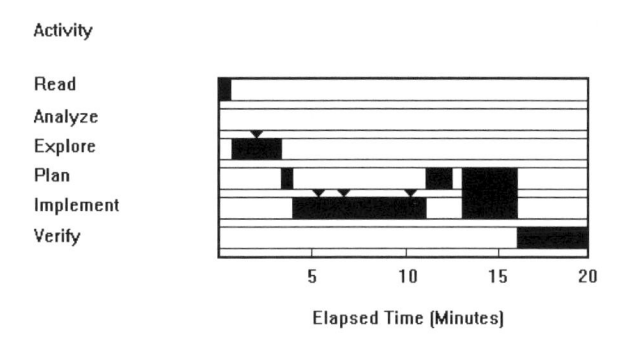

## An overall model of mathematical problem solving

I have considered many of the separate elements of knowledge and skill which are important for good mathematical thinking and problem solving. But in considering problem solving in mathematics, it is not enough to consider general strategies in isolation to arrive at an overview of the process.

Having the techniques to solve a problem may not be enough to enable a student to actually solve it. Even if the student has sufficient background knowledge and understanding of mathematical concepts, they may still fail to reach the solution to a problem. Clearly other factors also influence the problem-solving process (Lester *et al.* 1994). Lester and Charles (1982) suggest that three interacting sets of factors are at play, which all influence the problem-solving process.

1. Experience factors, both environmental and personal,

2. Affective factors, such as interest, motivation, pressure, anxiety and so on,
3. Cognitive factors, such as reading ability, reasoning ability, computational skills, and so on.

Experience factors could include such things as: age, previous mathematical experience and familiarity with the context within which the problem is set. The affective factors, over which the teacher or the student may have little control, are such as: stress, anxiety, pressure to perform and so on. While these two factors are interconnected they are also acted upon by the cognitive factors, creating a complex arena within which the problem-solving activity must take place. Thus, problem-solving can not merely be seen as a route to a desired solution, but as a complex and active process.

In his review of recent research Schoenfeld (1992) identified five psychological components or dimensions that interact and work together in mathematical problem solving. These are

- The knowledge base (facts, skills and conceptual structures)
- Problem solving strategies
- Monitoring and control (i.e. metacognitive processes)
- Beliefs and affects (appreciation and attitudes)
- Practices (knowledge of real-world contexts, experience in problem solving)

*Figure 7.8: The structure of memory and mathematical problem solving*

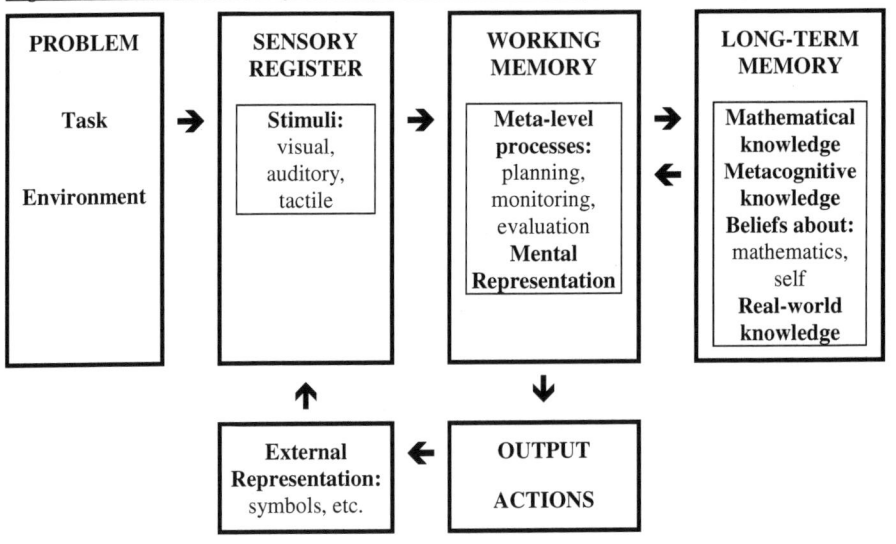

| PROBLEM<br><br>Task<br><br>Environment | SENSORY REGISTER<br><br>Stimuli:<br>visual,<br>auditory,<br>tactile | WORKING MEMORY<br><br>Meta-level processes:<br>planning,<br>monitoring,<br>evaluation<br>**Mental Representation** | LONG-TERM MEMORY<br><br>**Mathematical knowledge**<br>**Metacognitive knowledge**<br>**Beliefs about:**<br>mathematics,<br>self<br>**Real-world knowledge** |
|---|---|---|---|

External Representation: symbols, etc. ← OUTPUT ACTIONS

These are all issues that have come up separately in the above discussion. Schoenfeld interrelates these five psychological components in the following Figure

7.8, which shows a cognitive model which draws on the structure of human memory. It is based on the work of E. Silver (in Schoenfeld 1987).

Schoenfeld's account is based on the idea that much of the work in mathematical problem solving goes on in working memory (also known as short term memory). First of all, the problem is presented as a task in a given environment. Perceptions of this are fed to the sensory registers as visual, auditory, or tactile stimuli (i.e. we perceive the problem through our sight, hearing, etc.). These stimulate working memory to construct or recall and elaborate mental representations and initiate meta-level processes. Working memory is where mental representations of the problem situation (or the part of it current being worked on) are constructed and transformed. Of course most mathematical problem solving draws heavily upon external representations too, so there is output in the form of speech, writing, symbolism, diagrams, arrangements of objects or other forms of representation which are used and worked with during the process of problem solving. Perceptions of these are fed in through the sensory registers to refresh or update working memory representations.

In working memory there are also meta-level processes involved. These include planning ways of solving the problem, i.e., the means of transforming the problem representation towards the desired goal state, and deciding which processes or strategies or other items such as knowledge may prove useful. They also include monitoring mental activity, the progress made towards a solution, and keeping track of the effort involved. They involve the evaluation of plans, progress towards solutions, achieved solution states, and so on.

Working memory represents the active area of problem solving, of conscious thought, where feelings and attitudes are experienced. However what supports it is the content of long term memory. This includes mathematical knowledge, real-world knowledge, metacognitive knowledge and beliefs about mathematics and the self (as well as the knowledge and beliefs almost everything else that goes to make up a person).

Mathematical knowledge, from a psychological perspective, includes facts, skills, and conceptual structures. It also includes knowledge of past problems and problem solving approaches acquired through practical experience of school mathematics. There is also so-called 'real-world' knowledge, the knowledge acquired from working in contexts and practices beyond the mathematics classroom. This includes all the incidental knowledge needed to solve applied mathematical problems, to interpret language, to make sense of the social context and environment in which the task is set, and so on.

Metacognitive knowledge includes knowledge of the general strategies of mathematical problem solving, and knowledge of domain or topic specific strategies, such as successful means of transforming problem representations in the topic area towards solutions. It includes knowledge acquired through practical experience of monitoring mental activity, progressing towards solution, effort calculations, evaluation of plans, and other metacognitive activity. Lastly, there are beliefs about mathematics and about the self with regard to mathematics. These include conceptions of the nature of mathematics and of mathematical activity, as well as personal attitudes to mathematics and beliefs about the self as a learner of mathematics and as a mathematical problem solver.

Thus the overall cognitive model of problem solving shown in Figure 7.8 brings together all the elements of mathematical thinking discussed above and shows how they function as part of a whole. This is what is sometimes called an 'information processing' model of cognition, and the question may arise whether this fits with constructivism. I offer the model as one way of putting together the elements of thought described above. It does not presuppose either that ready made knowledge comes in via the senses (only perceptual stimuli) or that the representations created through the senses accurately reflect the world. Furthermore, it does not offer a mechanism for how the different elements listed interact in a mechanical or causal way. So although it is an information processing model it is quite consistent with a constructivist view of learning. Schoenfeld's account of human memory as used in problem solving thus draws together virtually all of the issues treated in this course.

There is a strong analogy between this model and the model of mathematical thinking proposed by Charles (1985) shown in Figure 7.9. This has four main components, thinking processes, attitudes, knowledge and beliefs.

*Figure 7.9: The model of mathematical thinking proposed by Charles (1985)*

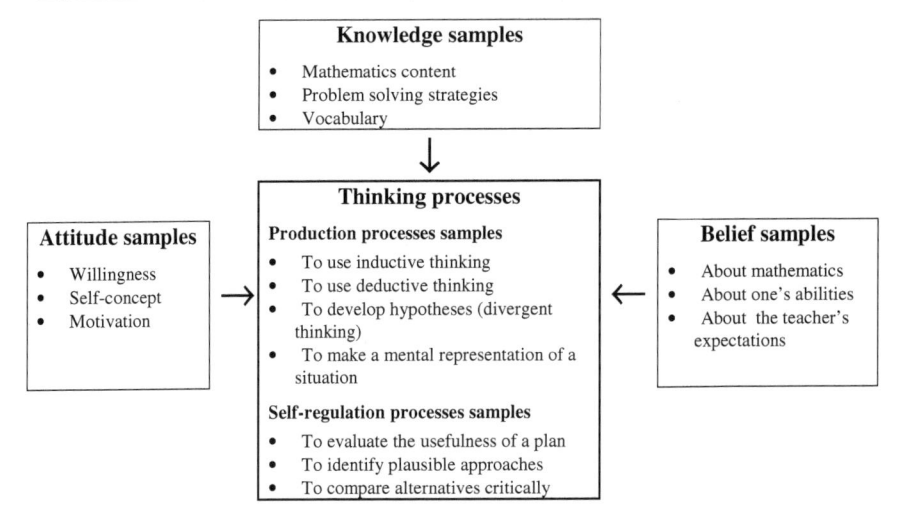

**Knowledge samples**
- Mathematics content
- Problem solving strategies
- Vocabulary

**Thinking processes**

Production processes samples
- To use inductive thinking
- To use deductive thinking
- To develop hypotheses (divergent thinking)
- To make a mental representation of a situation

Self-regulation processes samples
- To evaluate the usefulness of a plan
- To identify plausible approaches
- To compare alternatives critically

**Attitude samples**
- Willingness
- Self-concept
- Motivation

**Belief samples**
- About mathematics
- About one's abilities
- About the teacher's expectations

Charles proposes that mathematical thinking processes have two dimensions. First of all, there are the production processes, i.e., the thinking directly involved in carrying out problem solving, and he give examples of the use of inductive thinking (generalisation from examples to abstractions or patterns), deductive thinking, the development of hypotheses (e.g., speculation of patterns), and making a mental representation of a problem situation. Secondly, he distinguishes self-regulation processes, which are metacognitive processes. He gives as examples the evaluation of the usefulness of a plan, the identification of plausible approaches, and the critical comparison of alternative plans, approaches, etc.

These cognitive and metacognitive processes depend on the solver's knowledge, beliefs and attitudes. As examples of knowledge Charles lists mathematics content knowledge, knowledge of problem solving strategies, and knowledge of vocabulary. As examples of attitudes he lists pupil willingness, self-concept and motivation. Finally as relevant beliefs which impinge on mathematical thinking Charles lists beliefs about mathematics, about one's abilities, and about the teacher's expectations. All of these affect, he claims, pupils abilities and performance in problem solving and mathematical thinking in general.

## Teaching problem solving

One controversy in mathematics education is whether it is possible to teach problem solving strategies, and whether they actually help improve problem solving ability. It is known that explicit instruction in problem solving strategies without practice in problem solving is ineffective. The controversy is between those who argue that practical experience in problem solving should be supplemented with explicit instruction in problem solving strategies, and those who argue that such explicit instruction is ineffective, and that practical experience is all that counts. This latter claim, expressed strongly in this way, has already been refuted, because Schoenfeld's work demonstrates that interventions can raise the level of metacognitive activity and effectiveness in problem solving among students. There are a number of other experimental educational programmes which have a similar effect. For example, Reuven Feuerstein's (Sharron 1987) Instrumental Enrichment programme involves a set of activities designed to encourage general strategies, with the teacher explicitly bringing out and discussing the strategies used in the solution of problems. This has had remarkable results in raising the cognitive level of learner responses, especially with low attainers in mathematics (and across the curriculum). As in this case, an outcome of extended exposure to problems and strategy discussion often results in greatly enhanced performance by pupils. However, once again, this might be attributed to the development of metacognition and higher level thinking skills.

One approach involving the explicit teaching of problem solving strategies which is claimed to be effective in this area, is to concentrate on a single strategy at a time, such as drawing a picture, or making a table of results. To do this, a number of problems best solved by this strategy need to be collected and solved together. In the

1980s and early 1990s The USA journal *The Arithmetic Teacher* had a monthly feature Strategy Spotlight, that focused on the teaching of just one strategy each issue. In each case, an Arithmetic Teacher strategy spotlight section provided a set of problems focusing on one strategy, including the following:

- Guess and check (trial and error/improve)
- Draw a picture
- Make an organised list
- Make a table
- Find a pattern
- Write an equation
- Use logical reasoning

Experience is an essential contributor to problem solving ability, and this approach gives the student experience of a range of useful strategies (such as making an organised table of results) from which to choose, associated with a knowledge base of successful problem solving using that strategy.

What all sides in the controversy over whether to explicitly teach problem solving strategies agree on is that the key factor in encouraging problem solving, and the development of problem solving ability, in students is plenty of practice in solving non-routine problems. Practice leads to increased knowledge and confidence as the experience of successful problem solving builds up.

In summary, the dilemma of problem solving heuristics is whether to teach or not to teach heuristics. The argument *Anti* teaching heuristics is as follows:

- Doing problems, not studying heuristics is the way to teach and learn problem solving
- Research suggests only low attainers improve through strategy teaching (cf. R. Feuerstein's *Instrumental Enrichment* and Desforges and Cockburn 1987)
- Teaching heuristics routinises what should be creative, open, and thus turns what should be creativity into algorithm.
- Great mathematicians in history are self taught, and were not made or improved by instruction in heuristics.

The argument *Pro* teaching heuristics is that it:

- Allows and encourages a classroom focus on problem solving
- It raises awareness of the fact that there are many answers to problems and that there are many routes to an answer; and hence may help to develop learner's beliefs about and conceptions of the nature of mathematics
- Raises consciousness about strategies, and thus may give learners tools to use when stuck in problem solving

No-one suggests that the teaching of heuristics should replace problem solving experience, but that it should supplement it. What Schoenfeld and other's research suggests is that given extensive practical experience in problem solving, student performance can be enhanced by appropriate teacher interventions causing the student to become used to self-monitoring and controlling during problem solving.

Over half a century ago William Brownell (1942) developed a list of suggestions for teaching problem solving that are very apposite for today.

> To be most fruitful, practice in problem solving should not consist in repeated experiences in solving the same problems with the same techniques, but should consist of the solution of different problems by the same techniques and the application of different techniques to the same problems.

> A problem is not necessarily 'solved' because the correct response has been made. A problem is not truly solved unless the learner understands what he [or she] has done and knows why his [or her] actions were appropriate (Brownell 1942: 439)

> Instead of being 'protected' from error, the child should many times be exposed to error and be encouraged to detect and to demonstrate what is wrong and why (Brownell (1942: 440)

This third suggestion echoes the discussion in a previous chapter about the necessity of errors, as a concomitant of risk-taking and creative activity, in the learning of mathematics.

A number of researchers have identified important aspects of mathematical thinking that can (and have) significantly improved the outcomes in the teaching of problem solving. Mayer (1986) found that students forget or misrepresent relational aspects of problems (e.g., $X$ is twice $Y$ versus $X$ is 2 more than $Y$). This suggests that it is a good idea in teaching problem solving to focus directly on the relationships between variables.

Halpern (1992a) found that student problem solving is much more likely to be successful if students provide evidence that they understand the problem including being able to clearly state the goals of the problem. She also found that a common obstacle to generalising problem solving skills is the practice of teaching a solution principle (usually a domain specific one like the details of how to apply a particular statistical test) without requiring the student to determine the conditions under which it is applicable. (Notice that this echoes on a higher cognitive level the claims made for distributed practice of lower level skills and algorithms discussed in chapter 3 above).

Thomas and Grouws (1984) conducted an interesting related study of beginning college students' levels of cognitive operation. According to Jean Piaget's (1952, 1953, 1969) theory of Stages of Cognitive development, children's understanding passes through a number of distinct stages each with characteristic types of reasoning. In the *concrete operational stage* children develop the ability to transform their mental representations (the so-called concrete operations) from their experiences of changes in the real world. They also develop the ability to perform complex mental operations and learn that many concepts (e.g. number and measures) are conserved even when certain types of operations are performed on them. Only after a long period of being able to manipulate their mental images in the concrete operational stage are children ready for formal symbolic written work with understanding. Thus according to Piaget work on ratio and proportion, algebraic reasoning, logical deduction, etc., all require the child to have reached the *stage of formal operations*. This allows the learner to reason formally, abstractly and hypothetically.

Thomas and Grouws' study focused on beginning college students who performed at the concrete-operational level and not at the formal operational level on a Piagetian stages test. They randomly assigned students who performed at the concrete-operational level (and not at the formal operational level) on a Piagetian stages test to three group. Thus the students selected by Thomas and Grouws were slow in their cognitive development (although as they point out, in the USA only 25 to 40% of comparable students are at the formal operational level). Each of the three groups was given a different treatment.

1. The first control group (n=12) simply played draughts (checkers) with no interventions. The purpose of this group was to see if involvement in a research study was enough to have any impact.
2. The second control group (n=13) played Mastermind, the well known commercially available game in which one player has to guess the selection of coloured pegs mad by the other on the basis on information such as number of correct peg colours, number of pegs in correct positions of pegs, over a series of moves. In this group there were no researcher interventions.
3. The experimental group (n=14) also played Mastermind, but the researcher intervened with metacognitive-type questions, including: "Why do you think this colour is (or is not) in the code?", What did you learn from the last move?", Can you think of a way to get more information from your next move?", etc.

Each group spent an hour per week for 4 weeks engaged in the activities. They were tested both before and after on a cognitive development test (maximum score possible is 26). The results are shown in Table 7.10.

*Table 7.10: Results of treatments on cognitive development test.*

| Treatment | Pre-test scores | Post-test scores | Gain |
|---|---|---|---|
| Control group 1: Draughts | 12.4 | 13.8 | 1.4 (11%) |
| Control group 2: Mastermind without interventions | 11.9 | 14.6 | 2.7 (23%) |
| Experimental group: Mastermind with interventions | 11.4 | 16.1 | 4.7 (41%) |

The greatest cognitive gains occurred in the experimental group, who had the lowest mean pre-test score (not statistically significant) and by far the highest post-test scores and gain (significant at the 0.05 level). Qualitatively, in the experimental group 3 subjects (21% of group) changed from concrete operational stage to formal operational stage, 8 subjects (57% of group) changed from concrete operational stage to transitional stage (midway between concrete operational and formal stage) and the remaining 3 subjects were unchanged in terms of their Piagetian level.

In contrast, none of the control group 2 subjects changed from concrete operational stage to formal operational stage, although 6 subjects (50% of group) changed from concrete operational stage to transitional stage.

Thus it appears that guided interventions in the form of metacognitive questioning, not dissimilar to Schoenfeld's interventions described above, not only can help to enhance many of the concrete operational students' thinking strategies. It appears that they also have a transfer effect, in which the gains are transferred to levels of cognitive functioning/development more generally. This outcome echoes that reported above (Sharron 1987, Cockburn 1987), namely that teaching directed at improving learner strategies is effective with below average learners.

## Mathematical Investigation

The geographical metaphor for problem solving (Figure 5.1 above) is also applied to the process of mathematical investigation. "The emphasis is on exploring a piece of mathematics in all directions. The journey, not the destination, is the goal." (Pirie 1987: 2). Here the emphasis is on the exploration of an unknown land, rather than a journey to a specific goal. Thus whilst the process of mathematical problem solving is described as convergent, mathematical investigations are divergent (HMI 1985).

What is clear from this account is that problem solving in this sense involves attacking a set or given problem solving. This can be contrasted with what has become known n Britain as open ended investigatory mathematical work. This is broader than problem solving. Bell et al. (1983) propose a model of the process aspects of mathematics; the processes of investigation, with four phases. These are problem formulating, problem solving, verifying, integration. "Here the term 'investigation' is used in an attempt to embrace the whole variety of means of acquiring knowledge." (Bell *et al.* 1983: 207) They suggest that mathematical investigation is a special form, with its own characteristic components of abstracting, representing, modelling, generalising, proving, and symbolising. This approach has the virtue of specifying a number of mental processes involved in mathematical investigation (and problem solving). Whilst other authors, such as Polya (1945) include many of the components of the model as processes of problem solving, the central difference is the inclusion of problem formulation or problem posing, which precedes problem solving.

Another way in which mathematical investigation and problem solving are contrasted is that the former is a divergent activity while the latter is convergent. Problem solving converges on the desired given answer. Mathematical investigation starts in a given area of enquiry, then diverges according to the self-posed problems which are explored.

**Autonomy and control**

From the perspective of the teaching and learning of mathematics, an important feature of problem solving and investigatory mathematics is the degree of direct guidance and instruction and autonomy accorded to the student. These can be illustrated by contrasting the roles of the teacher and learner in different the teaching and learning approaches. This is shown in the following Table 7.11 (adapted from Ernest 1984) which also indicates the mathematical processes involved.

*Table 7.11: Contrasting teacher and learner roles in different the teaching approaches*

| Teaching Approach | Role of the teacher | Role of the learner | Processes |
|---|---|---|---|
| *Direct instruction (Exposition)* | Explicitly state knowledge. Provide exercises for application | Apply given knowledge to exercises | Application of facts, skills, and concepts |
| *Guided discovery* | Present rule implicitly in example-sequence | Infer rule implicit in examples | Generalisation, induction of pattern |
| *Problem solving* | Present problem. Leave method open | Attempt to solve problem by own method | Problem solving strategies |
| *Investigatory mathematics* | Present area of investigation or vet student choice | Choose questioned for investigation within topic. Explore freely | Problem posing and problem solving strategies |

This table indicates how as the teaching approach becomes less controlling, as it progresses from direct instruction to investigatory mathematics, so the role of the teacher changes, allowing more learner control over mathematical methods and processes, and finally, over the content itself. In this transition, the mathematical processes progress from the application of facts, skills, and concepts, to a limited repertoire of problem solving strategies including generalisation and the induction of pattern, to the full range of problem solving strategies, and finally adding problem posing and processes as well.

The shift involves more than mathematical processes. It also involves a shift in power with the teacher relinquishing control over the answers, over the methods applied by the learners, and over the choice of content of the lesson. The learners gain control over the solution methods they apply, and then finally over the content itself. The shift to a more inquiry orientated approach involves increased learner autonomy and self-regulation, the kind of outcomes achieved by successful meta-cognitive instruction.

# CHAPTER 8
## THE AFFECTIVE DOMAIN

In chapter 2 I distinguished, following Bloom *et al.* (1956), three domains representing psychological functioning and learning objectives. These are 1. The cognitive domain, concerned with knowledge, understanding, reasoning and intellectual functions; 2. the affective domain, concerning attitudes, feelings, and values; and 3. the psycho-motor domain, concerning physical skills and dexterity and perceptual skills.

Unlike the learning objectives discussed in earlier chapters which were primarily cognitive, attitudes and appreciation concern the affective domain. Just as in everyday speech 'affect' is cause and 'effect' is outcome, so too the affective domain is that area of causes internal to a person which drives their behaviours.

Bloom's Taxonomy of the Affective Domain (Krathwohl, Bloom *et al.* 1964) has five levels of ascending complexity and commitment. They are:
1. Receiving - Attending
2. Responding
3. Valuing
4. Organization (conceptualization of value; organization of value system)
5. Characterization (Characterization of values into consistent system).

In modern psychological usage the affective domain has been expanded beyond Krathwohl and Bloom's characterisation to encompass attitudes, feelings, emotions, beliefs and values.

In the present context it may be taken to include:
- Attitudes to mathematics: confidence, anxiety, liking maths (as a whole, or with reference to specific topics, schemes, and approaches).
- Beliefs about maths: beliefs about self and maths, about the nature of maths, about the nature of teaching and learning maths
- Appreciation of maths
- Perception of mathematics classroom climate (B. Fraser, R. L. Burden)
- Other aspects of the affective domain: values, feelings

In this chapter I shall focus only on the most significant aspects of affective domain as it concerns the psychology of learning mathematics, namely attitudes, beliefs and the appreciation of mathematics, with a few words about emotions and mathematics.

## ATTITUDES TO MATHEMATICS

The importance of attitudes to mathematics is widely recognised, because of a range of significant outcomes that can be associated with them. These include the following:

- ACHIEVEMENT – attitudes can have an influence on achievement in mathematics;
- PROBLEM SOLVING – is one of the main objectives of learning mathematics and attitudes contribute to persistence in problem solving;
- CHOICES – many students are opting out of maths, especially but not only, girls and women;
- ANXIETY – anxiety towards mathematics is debilitating both in education and throughout adult life;
- TEACHING – attitudes can be communicated by the teacher; whose view of maths can also be communicated.

Attitudes to mathematics and its teaching are important contributors to a teacher's make-up and approach, because of the effect they can have on a child's attitudes to mathematics and its learning (Aiken 1970). As is well known, affective factors generally do have a powerful influence on learning (Evans 1965, Khan and Weiss 1973). Researchers have found a significant correlation between teacher attitude and student achievement in mathematics (Begle 1979, Bishop and Nickson 1983, Schofield 1981), although it is often found to be weak (Begle 1979, Bell et al. 1983).

Over the years, a great deal of research has been undertaken into attitudes to mathematics. Significant surveys are given in Bell et al. (1983) and McLeod (1992). McLeod analyses the affective domain in mathematics to include attitudes, emotions, and beliefs. Beliefs are considered in the last section of this chapter. Emotions are 'hot' affect: feelings intensely experienced. These include frustration, joy, aesthetic appreciation (i.e., beauty). Emotions are often short lived and unstable, and have not been very deeply researched because of the difficulties involved. However attitudes toward some aspects of mathematics can generate strong emotions when that aspect is experienced, and so emotions have thus been studied indirectly, as products of attitudes.

Attitudes are 'cool' affect; they are enduring orientations or dispositions. Attitudes can be held towards mathematics, and towards the self as a student of mathematics, although in this latter case they are hard to separate from beliefs. Attitudes to mathematics are well entrenched and long-enduring dispositions towards mathematics. Thus, according to (Bishop and Nickson 1983: 20):

> As a result of the kind of mathematical experience they will have had at primary level and, more particularly, their achievement or lack of it with respect to the subject, attitudes to it are likely to be entrenched by the time they enter secondary school.

Attitudes mostly have two components: first they are about something (e.g., enjoyment of mathematics), secondly they have a direction or charge (positive or negative attitude towards something). McLeod (1992) defines attitudes as "affective responses that involve positive or negative feelings of moderate intensity and reasonable stability" (p.581), such as liking/disliking, interesting or boring etc.

In attitudinal research, researchers have distinguished up to a dozen factors, including, most significantly, the following:

1. Liking of mathematics (and its negative pole: dislike of mathematics). E.g., mathematics is 'interesting', 'exciting' or 'rewarding'. Enjoyment of and enthusiasm to do mathematical activities more or less fit under this construct (concept). e.g., mathematics is enjoyable or boring.
2. Confidence in one's own mathematical ability (and its negative pole lack of confidence). e.g., 'mathematics is easy', 'mathematics is difficult', 'I am good at mathematics', 'I am unsure about my own ability in mathematics'.
3. Anxiety towards mathematics: mathematics is related to negative feelings and experiences such as frustration, fear, pain or unpleasant situations.
4. Importance, utility or usefulness of mathematics, e.g., mathematics is necessary, important useful or irrelevant to modern life and work (a negative version).

| *Activity* |
|---|
| Sort out the attitude statements in Table 8.1: according to what attitude component is involved and whether each one is in positive or negative form. |

| *Assignment Suggestion* |
|---|
| Investigate student attitudes to mathematics, using the statements in Table 8.1as the basis for a questionnaire. |
| A study correlating attitudes to mathematics and achievement would be an interesting dissertation study E.g. testing the claim that a high attaining students in low sets have more positive attitudes to mathematics than low attaining students in high sets. |

*Table 8.1: Attitude to mathematics statements*

| Attitude to mathematics statements |
|---|
| 1. Maths is fascinating and fun |
| 2. It makes me nervous even to think about doing a maths problem |
| 3. I really like maths |
| 4. I avoid maths because I am not very good with figures |
| 5. I feel a sense of insecurity when attempting maths |
| 6. The feeling that I have about maths is a bad one |
| 7. I do not like maths; it scares me to have to study it |
| 8. Maths tests make me very nervous |
| 9. I enjoy most maths work |
| 10. Given the choice I would have nothing to do with maths |
| 11. I have always been confident about maths |
| 12. I feel at ease in mathematics, and I enjoy it |
| 13. Having to work out maths problems makes me nervous |
| 14. I enjoy solving maths puzzles |
| 15. Maths is a subject which I find difficult |
| 16. I am fairly confident about my mathematical ability |
| 17. At times I am fearful of maths |
| 18. I have never liked maths |
| 19. I don't feel sure of myself in maths |
| 20. I am less confident about maths than about other subjects |
| 21. I'm not very good at maths |
| 22. I find maths very interesting |
| 23. I'm afraid of maths |
| 24. Anything to do with maths makes me feel anxious |
| 25. Maths is a subject I find easy |
| 26. I have never been confident in maths |
| 27. I like mathematics better than most other subjects |
| 28. Maths homework has always worried me |

**Research on attitudes towards mathematics**

Cockcroft (1982: par. 205) reports the findings of the Assessment of Performance Unit as follows:

> Attitudes are derived from teachers' attitudes ... and to an extent from parents' attitudes (though this correlation is fairly low). Attitude to mathematics is correlated with attitudes to school as a whole (which is fairly consistent across the subjects) and with the peer-groups' attitude (a group attitude tends to become established).

Likewise, From the APU survey in 1985, in Costello found that:

The most accurate answer to the question of whether primary school children enjoy mathematics is 'sometimes'. It goes on to say that "During the secondary years, self reported attitudes to mathematics deteriorate amongst all groups of pupils... Attitudes, however, decline in a differentiated way. The acceptance that mathematics is useful, although weakened in secondary schools, stands up relatively well, whereas pupils become much more likely to dislike mathematics and describe it as hard. (Costello 1991: 123)

Since the nineteen seventies, extensive reviews of research on attitudes towards mathematics have been carried out (e.g., Aiken 1970, 1976, Kulm 1980, Leder 1987, McLeod 1992, 1994, Ma and Kishor 1997). These reviews show that studies on attitudes have been focused mainly on students' attitudes towards mathematics as a school subject. Research on student attitudes towards mathematics has also been conducted within a number of large scale evaluation studies, for example, the national study on pupils' attitudes towards mathematics by Assessment of Performance Unit (1988, 1991) and international comparison studies such as FIMS (Husen 1967), SIMS (Travers and Westbury 1989) and TIMSS (Keys *et al.* 1996a, 1996b and Harris *et al.* 1997). These studies often show that there is a tendency for attitude scores (especially on enjoyment and confidence subscales) to decline as students move from elementary to secondary school levels. Some major results illustrated from the APU surveys in Britain (1988 and 1991) are:

(a) 75% of both age groups (11-year-olds and 15-year-olds) perceived mathematics as useful.

(b) About one third of pupils from both age groups rate mathematics as difficult to some degree. Among the 11-year-old, 25 % of them are not sure whether mathematics is difficult or not.

(c) Comments made by the 11-year-olds suggest that enjoyment is an important factor on their attitudes towards mathematics, but pupils of aged 15 mentioned enjoyment far less frequently.

**Anxiety towards mathematics**

Anxiety towards mathematics is associated with the negative pole of two other attitudinal factors, i.e. strong dislike of, and extreme lack of confidence towards mathematics (although there is also test-anxiety in mathematics – i.e., fear of being tested). Mathephobia, fear or hatred of mathematics is another term for mathematics anxiety. Anxiety towards mathematics not very common among school-aged children (i.e., normally only a few percent will suffer from it), although it is significant amongst some adults (Buxton 1981, Maxwell 1989). Mathephobia was first identified in the literature in Gough (1954).

However, in her researches on children's attitudes to mathematics Hoyles (1982: 368) found that "anxiety, feelings of inadequacy and feelings of shame were quite common features" related to the 14 year-old pupils' bad experience of mathematics learning.

There is a large research literature concerning 'maths anxiety', especially among college students, e.g., Battista (1986), Betz (1978), Dew *et al.* (1983), Hendel (1980), Llabre and Suarez (1985), Morris (1981), Resnick *et al.* (1982), Richardson and Woolfolk (1980), Rounds and Hendel (1980), Tobias (1980, 1987). However most of these studies reflect the fact that many college students in the USA need to study mathematics as a requirement for graduating even though it is not their chosen subject of study.

Hembree (1990) conducted a meta-analysis of 151 studies researching mathematics anxiety. Not surprisingly he found that it correlates negatively with mathematics attainment, and is directly bound to avoidance of the subject. He also found that females display higher levels of mathematics anxiety than males.

Hodges (1983) argues that ensuring that there is match between teaching styles a learner's learning style can help prevent mathephobia. She presents a questionnaire to determine levels of mathephobia and refers to Dunn and Dunn's (1978) learning style inventory.

The most well known and widely used measure of mathematics anxiety in the research literature is the Mathematics Anxiety Rating Scale (MARS) (Suinn *et al.* 1972)

## GENDER DIFFERENCES IN ATTITUDES TOWARDS MATHEMATICS

Another important aspect of mathematics and attitudes to mathematics concerns gender issues. Not only do secondary school girls score lower on measure of attitudes to mathematics than boys; they also opt far less to pursue its study at GCSE 'A' level.

Results from the APU (1988 and 1991) surveys in Britain showed that in a comparison between genders, for both age groups 11 and 15, boys tended to display greater confidence than the girls in their mathematical abilities and performance. There is also a higher proportion of boys than girls who find mathematics interesting or enjoyable, and who appreciated the usefulness and value of mathematics.

There is a large body of research on mathematics attitudes as related to gender differences (Fennema and Sherman 1976, 1978, Eccles *et al.* 1993, Vanayan *et al.* 1997). The results of these studies consistently support the claim that there is an observed gender difference in mathematics attitudes as early as Grade 3 (Vanayan *et al.* 1997). Even though no gender differences were observed on many of the attitudinal variables such as 'liking' and 'enjoyment' (Steinback and Gwizdala 1995, Vanayan *et al.* 1997), more boys than girls reported that they are good at mathematics and boys are likely to believe that they are more competent than girls do. In the high school grades, more boys than girls tend to perceive mathematics as useful (Fennema and Sherman 1978). In short, boys tend to exhibit more positive attitudes towards mathematics than girls do and this disparity tends to increase with age.

Research on gender and attitudes reveals differences in attributions for success at mathematics (see Burton 1986). Table 8.2: indicates the commonly observed patterns of attributions for success and lack of success at mathematics by gender.

*Table 8.2: Attributions for success & lack of success at mathematics by gender*

|  | **Attributions for maths success** | **Attributions for lack of success** |
|---|---|---|
| **BOYS** | Skill, Ability | Bad luck, Lack of effort |
| **GIRLS** | God luck, Effort | Lack of Skill and Ability |

More typically boys attribute their success at mathematics to stable and intrinsic causes such as their skill and ability; and their failures to extrinsic and unstable causes such bad luck, or a lack of effort. The girls, however typically attribute their success at maths to extrinsic and unstable causes such good luck, or to study effort, and their failures to stable and intrinsic causes such as their lack of skill and ability. This topic is treated in greater depth in the advanced course module on Gender and Mathematics.

## ATTITUDES AND ACHIEVEMENT IN MATHEMATICS

"Abilities, preferences, attitudes and motivation all contribute to making some pupils more successful than others." (Orton 1992: 107). As this quotation indicates, perhaps the most significant outcome of attitudes to mathematics is their relationship to achievement.

> There is a common and reasonable belief that positive attitudes, particularly liking for, and interest in, mathematics, lead to greater effort and in turn to higher achievement... What is more clear is the strong relationship between attitude and the choice of mathematics for further study... Positive attitudes, however, can be regarded as valid objectives of mathematics education in their own right. Affective learning outcomes - such as enjoyment, enthusiasm, fascination, appreciation - may be taken into account alongside the more cognitive aspects of learning mathematics which are measured in terms of achievement. (Costello 1991: 122)

A significant correlation between positive attitude to mathematics and achievement has been found (see, e.g., Bell *et al.* 1983, McLeod 1992). Undoubtedly positive attitudes matter a great deal in the mathematics teaching and learning. However the relationship between attitudes and achievement is not simple. For example, studies at Exeter have shown that a high attaining student in a low set may have a more positive attitude to mathematics than a relatively low attaining student in a higher mathematics set, *irrespective of the fact that the second student's overall attainment is higher in absolute terms.*

Another muddying factor is that of causal link. If higher attaining students have more positive attitudes to mathematics, which is the cause? Do positive attitudes to mathematics cause high attainment, or does high attainment cause positive attitudes? Correlation studies alone do not provide the means of answering these questions. (see the self-reinforcing cycles in Figures 8.2 and 8.3 below).

Overall, there has also been much research on mathematics attitudes investigating the relationship between attitudes and achievement (Quinn and Jadav 1987, Minato and Kamada 1996); and the possibly correlated factors that influence students' attitudes towards mathematics (Reynolds and Walberg 1992). Both narrative literature reviews (Aiken 1970, 1976, Neale 1969, McLeod 1992) and meta-analysis (Ma and Kishor 1997) on attitudes studies indicate that there is no consistent findings supporting the relationship between attitudes towards mathematics and achievement in mathematics. This relationship was found to be statistically significant but not strong enough for educational practice (Aiken 1970, 1976, Neale 1969, Ma and Kishor 1997). While others found them to "interact with each other in a complex and unpredictable ways" (McLeod 1992: 582). Although most studies claim that positive attitudes to mathematics might lead to better mathematics achievement but this causal relationship is normally weak and not confirmed by research evidence. Therefore, Ma and Kishor (1997) point out that most probably there is a potential mediating variable between these two factors such as motivation or mathematics participation in class.

In general, the possibly correlated factors claimed to have strong link with students' attitudes to mathematics are motivation (Reynolds and Walberg 1992), parents' attitudes towards mathematics (Cain-Caston 1986), home environment (Cai, Moyer and Wang 1997), cultural factors (Huang and Waxman (1997), and teachers' instructional quality as perceived by students (Reynolds and Walberg 1992). These findings, however, are still far from conclusive because not all studies produce consistent evidence.

The Assessment of Performance Unit (1985) investigated the correlation of attitudes with achievement scores in mathematics. Their results are summarised in Table.

*Table 8.3: Correlation of attitudes with achievement scores*

|        | Enjoyment | Utility | Difficulty (Perceived) |
|--------|-----------|---------|------------------------|
| Age 11 | 0.21      | 0.43    | 0.42                   |
| Age 15 | 0.13      | 0.20    | 0.40                   |

(source Assessment of Performance Unit 1985: 528)

Although because of the large sample size tested all results were statistically significant at 0.1% level, the correlations were not very high. At Age 11 the correlation of enjoyment with achievement was only 0.21, and by Age 15 this had reduced to 0.13. Perceived difficulty of mathematics was better correlated with achievement, and was at about the 0.4 mark for both 11 and 15 years olds.

Lorenz (1982) proposes a model of the relationship between attitudes and achievement in mathematics. He suggests that at the heart of affect is the child's mathematical self concept. This, coupled with input from the learner's general (i.e., trait or personality) anxiety feeds into the child's mathematical anxieties. The child's mathematical self concept also feeds into their motivation and effort calculations when faced with a particular task or assessment. These three factors (mathematics anxiety, motivation, effort calculation) lead to the child's persistence which directly affects their mathematical achievements and test results (supported by cognitive factors such as knowledge and skill). In addition, these three factors are manifested in the child's willingness to respond and interact in mathematics. Lastly, according to Lorenz, the child's mathematical self concept also causes their attitude to mathematics, which contributes to their school attitude. This factors and their proposed links are shown in Figure 8.1.

*Figure 8.1: Lorenz's model of the relationship between attitudes and achievement*

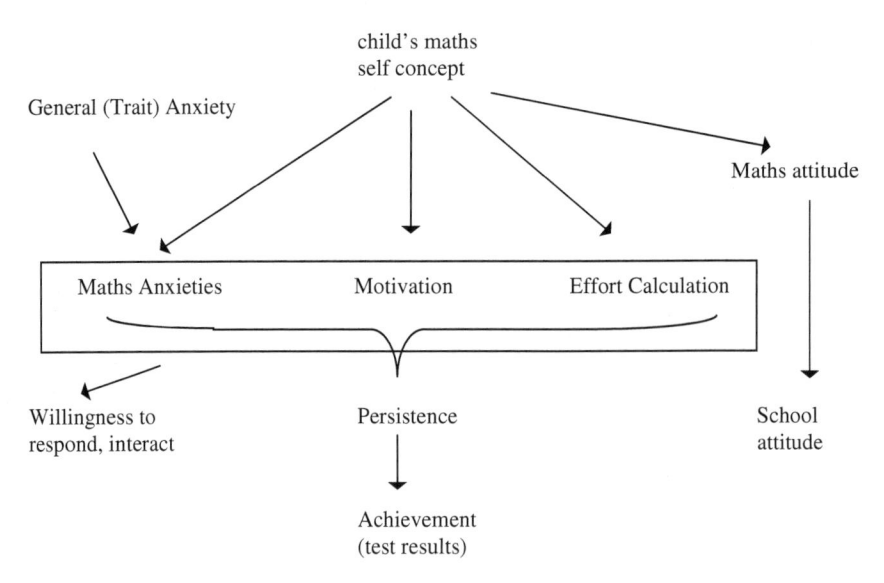

According to the Lorenz model, a learner's mathematical attitudes including the learner's mathematical self concept and motivation, combined with other factors, cause the learner's persistence which itself is a significant cause of their mathematical achievement and test results. These causal links form a cycle which reinforce the learner's attitudes to mathematics. Thus failure at mathematical tasks may not only be the outcome of poor attitudes, motivation and mathematical self-concept, but may also contribute to and reinforce negative attitudes.

> In mathematics more than any other subject there is the possibility that they [learners] will experience absolute failure at the tasks they are given. (Haylock 1991: 35)

Repeated failure in mathematics is thus likely to affect the attitudes and motivation of a learner. Haylock (1991) claims that "Repeated failure in mathematics increases the level of anxiety, because of fear of further failure; then the anxiety inhibits the child's performance and leads to further failure."

This is described by Ernest (1985) as a self-reinforcing, vicious 'failure cycle' and the effect it has on the child is to erode their confidence, which in turn leads to a lack of motivation and reduced concentration and persistence.

*Figure 8.2: The Failure Cycle*

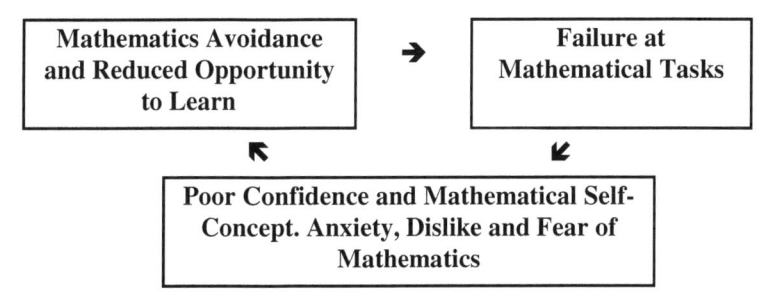

| Mathematics Avoidance and Reduced Opportunity to Learn | → | Failure at Mathematical Tasks |
|---|---|---|
| ↖ | | ↙ |
| | Poor Confidence and Mathematical Self-Concept. Anxiety, Dislike and Fear of Mathematics | |

Because the student experiences failure at mathematics, the student loses confidence in his or her mathematical abilities, may feel anxious about mathematics, and may fear or dislike the subject. As a consequence of these negative reactions the student lacks confidence and persistence, and may avoid mathematics, or keep to safe, easy and familiar parts. The resulting lack of practice and lack of extension of knowledge and skills represents a reduced opportunity to learn, and leads to further failure. This reinforces fear and dislike of the subject, and further damages self-confidence, mathematical self-concept, and self-image in general. This is a self-perpetuating cycle of failure that is difficult to break out of, and can continue in a descending vicious spiral.

In contrast, the experience of success and positive attitudes can combine to form a virtuous cycle.

*Figure 8.3: The Success Cycle*

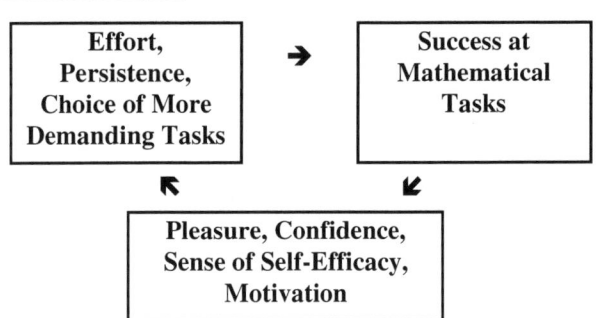

```
┌─────────────────┐         ┌─────────────────┐
│     Effort,     │         │   Success at    │
│   Persistence,  │   →     │  Mathematical   │
│  Choice of More │         │     Tasks       │
│ Demanding Tasks │         │                 │
└─────────────────┘         └─────────────────┘
          ↖                         ↙
         ┌─────────────────────────┐
         │  Pleasure, Confidence,  │
         │  Sense of Self-Efficacy,│
         │       Motivation        │
         └─────────────────────────┘
```

Once students are experiencing success at mathematical tasks they should begin to enjoy doing mathematics. The success and pleasure lead to confidence, and over a period of time, to improved mathematical self-concept and sense of self-efficacy. These positive responses improve motivation which leads to more serious commitment and effort, and persistence. The increased effort and work properly managed will give rise to more success, completing the success cycle. This should be an upward spiral. At first, the increased efforts will arise from the desire for external approval and the reassurance it brings: i.e., extrinsic motives. It is to be hoped that after a while intrinsic motivation will develop, enjoying the activities for their own sakes and the sense of achievement success brings.

These three models (Figures 8.1, 8.2 and 8.3) suggest mediating variables between attitudes to mathematics and mathematical achievement, such as motivation, effort and participation in mathematics class, reinforcing Ma and Kishor's (1997) conclusions mentioned above.

## ATTITUDES TO MATHEMATICAL TOPICS

There has been some research into children's attitudes to specific mathematical topics. The Assessment of Performance Unit's research with 11 year olds found that certain topics were thought easy by almost all, whereas the hardest topics were thought hard or very hard by more than 50% of the sample (Joffe and Foxman 1988).

*Table 8.4: 11 year olds attitudes to mathematical topics*

| EASIEST TOPICS | HARDEST TOPICS |
|---|---|
| *Thought easy by almost all* | *Thought hard/very hard by more than 50%* |
| Adding | Angles |
| Subtracting | Fractions |
| Money | % |
| Time | Volume |
| Multiplying | Area |
| Dividing | Factors |

Carpenter and colleagues (1980) reported the findings of the USA National Assessment of Educational Progress. They also describe the attitudes to mathematical topics, but of 9 year olds, as shown in Table 8.5 below.

*Table 8.5: 9 year olds attitudes to mathematical topics*

| TOPIC / ACTIVITY | LIKE / DISLIKE % | | EASY / HARD % | |
|---|---|---|---|---|
| learning about money | 68 | 7 | 52 | 9 |
| addition problems | 65 | 10 | 65 | 9 |
| subtraction problems | 47 | 20 | 59 | 12 |
| dividing | 50 | 18 | 39 | 16 |
| solving word problems | 45 | 18 | 44 | 18 |
| learning times tables | 56 | 24 | 39 | 23 |
| learning about shapes | 63 | 11 | 82 | 4 |

These results show wide variations in affective responses to mathematical topics. In the USA sample 56% like learning times tables and 24% find this easy. However 24% dislike learning times tables and 23% find this hard, the highest topic scores for disliking a topic and finding it difficult.

**MEANS OF ASSESSING ATTITUDES**

A number of means are available for investigating attitudes. However there is an intrinsic difficulty. Most means of assessment rely on learner (or other subject's) self-reports. This brings with it several methodological dangers. Learner self-reports may misrepresent the subject's attitudes, because

1. They may feel they ought to express certain attitudes which do not accurately reflect their attitudes (because of prevailing or group attitudes, to please the teacher or researcher, etc.),
2. They may not actually know what their attitudes are,
3. They may answer impulsively, making up answers on the spot, leading to poor reliability.

These weaknesses and difficulties must be borne in mind in any investigation of attitudes to mathematics.

The means available for such research include the following: questionnaires (both closed, using fixed responses, or open ended questions), semantic differential questionnaires, interviews with individuals or small groups of learners, and observations of behaviour. These are discussed with additional details below.

Questionnaires can be Likert Type (like Table 8.6 below, including a number of fixed responses such as Strongly Agree, Agree, Undecided, Disagree, Strongly Disagree (these can be scored 2,1,0,-1,-2, respectively).

Variations can include verbal items for subjects to respond to, mathematical items for them to rate, etc. Responses can be choice of fixed responses, as in Likert scales, and a variation with smiley faces has been used to make the questionnaire accessible to young children (☹, ☺, see Dunlap 1976). In this latter case, the questions are usually read out to the individuals or class to circumvent language difficulties.

In order to investigate a learner's own ideas one can use open questions, with verbal responses, e.g., My feelings about maths are.......... . However such results are harder to tabulate and analyse, but would be appropriate for interpretative research into attitudes, exploring children's own ideas, meanings and feelings about mathematics.

Another variation in questionnaires is the use of Semantic Differential questionnaires. Subjects mark a position on a continuum towards either pole:

Mathematics is:     Easy ............................................Hard
                            Interesting...................................Uninteresting

This indicates, with a minimum of writing, how learners attribute meanings to mathematics.

Interviews are another means of investigating attitudes. These allow individual's (or small groups of learners') attitudes to be probed and explored in depth, subject to some of the difficulties mentioned above.

Observation of behaviour where it is possible can provide reliable information on attitudes. Thus learner choices about further study, and their attendance on courses are indications of their commitment (although extraneous factors may also intrude, e.g., finance, home situation).

Leder (1985) and Michaels and Forsyth (1978) discuss some of the means of measuring attitudes towards mathematics, and provide some helpful pointers in choosing and designing instruments for this purpose.

## PERCEPTIONS, BELIEFS AND APPRECIATION OF MATHEMATICS

It is widely remarked in the mathematics education literature that student and teacher attitudes and perceptions of mathematics are important factors in learning (e.g., Ernest 1991a). Indeed, much of the research on girls and mathematics talks of the problems caused by the stereotypical perceptions of mathematics as a male domain (e.g., Walkerdine 1988, 1998). Such views are prevalent, especially in Western Anglophone countries. This area is only mentioned here, but is explored in depth in the advanced course module on Gender and Mathematics.

It might be said that every teacher of mathematics - perhaps every teacher - should ask him or herself the question 'what is mathematics?' at least once: Our view of the nature of mathematics affects the way we teach, and will affect the way the way the children we teach view mathematics. Yet it is all too rarely that we stand back and take a broader view of mathematics, let alone share this view with the children we teach.

While there is a large research literature on attitudes towards mathematics, research on beliefs of mathematics is only gaining attention more recently. In mathematics education, research on beliefs has become a major focus within two main areas of research, namely problem solving (see Schoenfeld 1989) and gender difference in mathematics (see Fennema et al. 1990), mentioned above. More recently, there is an increasing number of studies relating teachers' beliefs and their mathematics teaching in class (see the reviews in Pajeres 1992, Raymond 1993).

## PHILOSOPHY OF MATHEMATICS

Views concerning the nature of mathematics as a whole form the basis of the what is called the philosophy of mathematics, which is a branch of philosophy (see, e.g., Davis and Hersh 1980, Ernest 1991a, 1998a). Unlike the academic study of philosophy, teachers' philosophies, that is their views of the nature of mathematics, are not necessarily consciously held views. They may instead be implicitly held philosophies, which the teachers have not stopped to consider consciously, but are there all the same. There are many different viewpoints about mathematics. However a central distinction in views of the nature of mathematics is between two types of position. These are Dualist views of mathematics and non-Dualist or Relativist. The distinction is based on the work of William Perry (1970) which has been applied to views of mathematics by many people including Cooney (1988), Ernest (1991a) and Carré and Ernest (1993). Of course this distinction is simplistic, and if people's views can be located it will be along a continuum. Nevertheless, the distinction is an important one

The Dualist view is that the mathematics is a fixed and absolute set of truths and rules. Mathematics is exact and certain, cut and dried, right or wrong, and there is always a rule to follow in solving problems

The Relativist (non-Dualist) view is that mathematics is a dynamic, problem-driven and continually expanding field of human creation and invention, in which patterns are generated and then distilled into knowledge. This view places most emphasis on mathematical activity, the doing of mathematics, and it accepts that there are many ways of solving any problem in mathematics.

Some of the curriculum reform movements in mathematics have promoted views of mathematics that fit this classification. On the one hand, the Back-to-Basics movement which emphasises basic numeracy as knowledge of facts, rules and skills, with little regard for meaning or problem solving, can be regarded as promoting a Dualist view of mathematics. On the other hand, the emphasis on of problem solving and investigational work in mathematics, in the Cockcroft Report (1982) and the National Curriculum in Mathematics Non-Statutory Guidance (National Curriculum Council 1991) supports a non-Dualist view of mathematics. For example, the latter contrasts Closed mathematical tasks (single fixed answer) from those which are not-Closed (these have multiple answers, some times an unlimited number). In terms of the concepts of the previous chapter, a Dualist view of mathematics would not regard children posing their own problems as anything to be encouraged or tolerated. In contrast, a non-Dualist view of mathematics would regard such activity as potentially a valuable contribution to mathematical proficiency.

## VIEW OF MATHEMATICS QUESTIONNAIRE

The following (Table 8.6) is a questionnaire for identifying beliefs about mathematics, whether they tend to be Dualist or non-Dualist (Relativist).

*INSTRUCTIONS*
Please mark the box to show your opinion. The responses are short for: Strongly Agree, Agree, Undecided, Disagree, Strongly Disagree.

*Table 8.6: View of mathematics questionnaire*

| | | | | | |
|---|---|---|---|---|---|
| 1. Maths consists of a set of fixed, everlasting truths | SA | A | U | D | SD |
| 2. There are many ways of solving any problem in maths | SA | A | U | D | SD |
| 3. There is always a rule to follow in solving maths problems | SA | A | U | D | SD |
| 4. Some maths problems have many answers, some have none | SA | A | U | D | SD |
| 5. Learning maths is mainly remembering facts and rules | SA | A | U | D | SD |
| 6. Maths is basically doing sums | SA | A | U | D | SD |
| 7. Puzzles and investigations are not proper maths | SA | A | U | D | SD |
| 8. Maths is always changing and growing | SA | A | U | D | SD |
| 9. The procedures and methods in maths guarantee right answers | SA | A | U | D | SD |
| 10. There is only one correct way of solving any maths problem | SA | A | U | D | SD |
| 11. A person should not mind risking a mistake when trying to solve a maths problem | SA | A | U | D | SD |
| 12. Maths is always exact and certain | SA | A | U | D | SD |
| 13. Knowing how to solve a problem is even more important than getting the right answer in maths | SA | A | U | D | SD |
| 14. Exploring number patterns is not real maths | SA | A | U | D | SD |

**Scoring the view of mathematics questionnaire**

A Dualistic view of mathematics is asserted in items: 1, 3, 5, 6, 7, 9, 10, 12, 14. If these are scored: SA = 1, A = 2, U = 3, D = 4, SD = 5, then the mean score for these items (total/9) gives a measure of a Dualistic view of mathematics, the lower the stronger (neutral = 3).

Non-Dualistic view of mathematics is asserted in items: 2, 4, 8, 11, 13. If these are scored: SA = 5, A = 4, U = 3, D = 2, SD = 1, then the mean score for these items (total/5) gives a measure of a non-Dualistic view of mathematics, the higher the stronger (neutral = 3).

These mean scores can be added together to give a score between 10 (non-Dualistic view) and 2 (Dualistic view), with neutral at 6.

| Assignment Suggestion |
|---|
| Investigate student or teacher views of mathematics. This can be done by structured interviews, or by using a questionnaire, such as that given below. The results of the use of a similar questionnaire are given in Carré and Ernest (1993). |

## BELIEFS AND VIEWS OF MATHEMATICS

Closely related to studies on beliefs of mathematics are studies on mathematical myths. There is an increasing number of studies (see example, Mtetwa and Garofalo 1989, Frank 1990) investigating mathematical myths that are held by students, and preservice teachers in particular. As defined by Frank (1990: 10), a 'mathematics myth' refers to "a belief about mathematics that is (potentially) harmful to the person holding that belief because belief in a maths myths can result in false impression about how mathematics is done".

Kogelman and Warren (1978) identified twelve myths that are commonly related to those mathematics anxious and mathematics avoidance students. They are:

1. Some people have a maths mind and some don't.
2. Maths requires logic, not intuition.
3. You must always know how you got the answer.
4. There is a best way to do a maths problem.
5. Maths requires a good memory.
6. Maths is done by working intensely until the problem is solved.
7. Men are better in maths than women.
8. It's always important to get the answer exactly right.
9. Mathematicians do problems quickly in their heads.
10. Maths is not creative.
11. It is bad to count on your fingers.
12. There is a magic key to doing math.

When analysed further, the list shows that many of these myths are related to beliefs about the nature of mathematics and learning mathematics. Myths (1) and (7) suggest that there are gender differences in mathematical ability; myths (2), (8), (9) and (10) seem to imply that mathematics is a logical, rigid and hierarchical subject. While myths (3), (4), (11) and (12) suggest more of a dualistic view that there is a fixed way of getting the right answer, myths (5) and (6) indicate that memory and effort are important in doing mathematics.

Perhaps this is best summarised by Paulos (1992: 335) when he proposes that there are at least five "mathematics-moron myths" that need to be exploded by mathematics educators and teachers because they are as important as other educational issues such as curriculum reform and the use of technological tools. According to him, these five myths are:

1. Mathematics is computation;
2. Mathematics is a rigidly hierarchical subject;
3. Mathematics and narrative are disparate activities;
4. Mathematics is only for the few; and
5. Mathematics is numbing. (p.335)

These myths are also evidenced in Mtetwa and Garofalo's (1989) study. They investigated beliefs about mathematics held by students with difficulties with mathematics. They identified five myths that were commonly held by these pupils, which include, computation problems must be solved by using a step-by-step algorithm, and mathematics problems have only one correct answer. Perhaps holding these myths might have further discouraged these students from liking mathematics and as a result, they face difficulties in mathematics.

In short, these studies indicate a widely held mathematical myth common to both students and preservice teachers: there is always a right answer in mathematics. This myth is consistent with the Dualist view detailed above.

In considering research on student and teacher beliefs about mathematics it is important to distinguish between:

1.  teachers' views of mathematics as a discipline in its own right
2.  teachers' views of school mathematics
3.  students' views of school mathematics

It is widely argued that there is a connection between (1) and (2), although it is not a simple one, and that likewise there is a connection between (2) and (3). See Cooney (1988), Ernest (1991a) and Carré and Ernest (1993). It is not possible in the limited space here to go into all the complexities of these relationships. What can be said is that the connection is not a simple logical or causal one; it is rather, at best, a statistical correlation which is greatly influenced by contextual factors. The influence of the social context may, in fact, neutralise or overpower such influences (Ernest 1989, 1995).

## STUDENTS' VIEWS OF SCHOOL MATHEMATICS

Erlwanger (1973) carried out a case study of a confident, intelligent, and very articulate 12 year old called Benny. Benny studies an individualised scheme in the USA called *IPI Mathematics* (Individually Prescribed Instruction). Benny is successful, according to the built in checks of the IPI scheme, and yet he constructs incorrect rules for himself, and is wholly failing to understand mathematics in the way that is presumably intended. On the basis of his experiences he has constructed a view of mathematics as irrational "magic", and a "wild goose chase" in which he looks for clues to find the answers to match those in the official answer book. He has learnt that there are many right answers to each problem, but that only one is in the approved form required. His view of mathematics is that of an unrelated collection of rules ("in fractions we have 100 different kinds of rules"). The rules are arbitrary, in terms of reason, but have the sanction of authority. This is very much like the Dualist view of mathematics.

In a research study M. Preston found two views of mathematics in a sample of almost three thousand school children. The children were

A. Tending to see mathematics as an algorithmic, mechanical and somewhat stereotyped subject, or
B. Tending to see mathematics in an open ended, intuitive and heuristic setting.

This evidence confirms that children do construct powerful images of mathematics for themselves, on the basis of their learning experiences. Their views are very reminiscent of the philosophies of mathematics discussed in chapter 1. View (B) is like the problem solving philosophy of mathematics, whilst view (A) is closer to (but not identical to ) a Dualist view of mathematics. Preston found that the children's views depended a great deal upon the type of course or text they were following, and on factors like the number of different mathematics teacher they had, the children's gender, and so on. A modern course which treated mathematics in context was associated with view (B), whilst having more teachers of mathematics, and being a girl, tended to be associated with view (A). (See Preston 1975, also reported in Bell *et al*. 1983)

A number of other researchers have explore student view of mathematics. Kouba and McDonald (1987, 1991) compared the views of secondary and primary school students with regard to what they considered to be mathematics, and what they did not consider to be mathematics. They issued questionnaires to 451 secondary students and 1202 primary school students. They found that unlike adults, children do not view mathematics to be a well defined subject matter. With regard to the results with primary school students they said that the students largely identified mathematics with counting and number operation work. They also regard it as an exclusive domain, school based and isolated form other areas of study. In addition "For them, the domain of mathematics, while being narrow, is also not constant. Rather it is upwardly shifting. To many children when something becomes easy, it is no longer mathematics." (Kouba and McDonald 1987: 107)

They found that the lower secondary students (years 8 and 9) viewed mathematics as a broader domain, including probability, geometry and measurement (as well as arithmetic). However, they did so mostly where an explicit mathematical term, symbol, or format was used, and not when the concept of symmetry was involved, but not the term. Overall, and perhaps not surprisingly, both groups of students identified as maths that which they did in the maths lesson, and which had been identified there as maths. Thus the conclusion to be drawn is that pupil's classroom experiences are decisive in developing their views of mathematics.

The Assessment of Performance Unit (1985) conducted extensive investigations into perceptions of mathematics, as well as towards attitudes to it. They found that students distinguished mathematical topics as hard-easy and as useful-not useful, and that theses categories played a significant part in their overall view of mathematics. They also found that students tended uniformly to regard mathematics as a whole as both useful and important, reflecting a realistic perception of the weight that is attached to the subject in the modern world.

Using both quantitative (such as survey questionnaires) and qualitative methods (such as observation and extensive interviews), Frank (1988) investigated 27 mathematically high attaining middle school (i.e., early secondary school) students on their mathematical beliefs and how their beliefs might influence the way they solve problems. His findings can be summarised as follows. The students tended to believe that:

1. Mathematics is computation and followed that 'doing mathematics means following rules and learning mathematics is mostly memorisation'.
2. Mathematics problems should be quickly solvable in just a few steps.
3. The goal of doing mathematics is to obtain the 'right answers'
4. The role of mathematics students is to receive mathematical knowledge and to demonstrate that it has been received.
5. The role of the mathematics teacher is to transmit mathematical knowledge and to verify that students have received this knowledge. (p.33)

Likewise, Cesar (1995) asked 331 pupils from the 7$^{th}$ level in Lisbon schools for their images of mathematics as well as their reason of liking or disliking it. She found that most of them viewed mathematics as computation and some of them described mathematics as useful for daily or future life. The main difference between those who like mathematics and those who dislike mathematics, however, was that the former found mathematics interesting while the latter perceived mathematics as complex and difficult.

There are also many major national and international evaluation studies such as National Assessment of Education Progress (Brown *et al.* 1988), SIMS (Travers and Westbury 1989) and TIMSS (Keys *et al.* 1996a, 1996b and Harris *et al.* 1997) have included beliefs about mathematics as one of their variables. In general, these data suggest that students believed that mathematics is important, useful but difficult and mainly involve memorising rules.

Students' views of school mathematics probably depend on the sum of their classroom experiences of mathematics, including the quality of the relations they have with their teacher; institutionalized classroom competition; the extent of negative weight placed on errors; the degree of public humiliation experienced in consequence of failure; and other such factors which powerfully impact on the young learners self-esteem and self-concept as a learners of mathematics. Dualistic views of mathematics can be communicated by giving students a myriad of unrelated routine mathematical tasks which involve the application of memorized procedures, and by stressing that every task has a unique, fixed and objectively right answer, coupled with disapproval and criticism of any failure to achieve this answer. Unfortunately the experience many learners have during their years of schooling confirms the negative Dualistic image of mathematics as cold, absolute, inhuman and rejecting. Such an image is frequently associated with negative attitudes to mathematics. This is verified in studies of adults' attitudes and responses to mathematics, such as Buerk (1984) and Sewell (1981).

Thus my claim is that experiences in school mathematics form the basis for the conceptions, appreciation and images of mathematics constructed by learners, especially negative ones. Of course only a portion of classroom experiences or images of mathematics are negative. Just as a single bad experience can produce a negative image of mathematics, so too a single good experience can provide the basis for the development of a positive image of mathematics. Womack (1983) illustrates this when he describes his personal experience in which an interest in pursing a single mathematics problem in school led to success, teacher encouragement and the growth of a fascination with mathematics which resulted ultimately in his choice of a career in mathematics education.

This and other evidence suggests that children construct powerful stereotyped images of mathematics for themselves, apparently based on their classroom learning experiences. Any impact of massmedia or other popular images of mathematics are harder to discern, as students' classroom experiences seem to be the dominant influences.

## GENERATING POSITIVE BELIEFS ABOUT MATHEMATICS

One suggestion is that the history of mathematics can be introduced into mathematics teaching and learning to give it a more human face. Fauvel (1991) suggests that the use of the history of mathematics in teaching can have the following outcomes.

- Helps to increase motivation for learning
- Makes mathematics less frightening
- Pupils derive comfort from knowing they are not the only ones with problems
- Gives mathematics a human face
- Changes pupils' perceptions of mathematics

Although there is insufficient research on the relationship between the use of the history and positive attitudes to and useful perceptions of mathematics, a number of experimental teaching programmes use a historical approach and have reported positive outcomes (e.g., some Italian experiments described by Bartolini-Bussi 1991).

# CHAPTER 9

## THE CONTEXT OF LEARNING MATHEMATICS

It has long been recognised that the contexts of learning mathematics and the contexts of mathematical tasks play a vital role in the teaching and learning of mathematics. However, there is also an ambiguity in the term 'context'. For the social context of a task or learning situation means one thing, whereas the contextualisation of a task or a problem means something else, and both are referred to as the 'context of a task'. The former concerns the social location of the task, the modes of day-to-day activities surrounding it, the roles of persons involved. In contrast, the latter concerns the way the task is represented, the way it is dressed up as a problem with reference to objects, activities, etc., but typically communicated in written or pictorial form as a text-book problem.

'Real' problem solving is supposed to involve realistic problems as one meets them in actual work, leisure or other situations. In other words, contextual factors ('context' literally means 'with-text', the additional factors accompanying a text) such as references to everyday objects, persons and work practices, and diagrams and figures also referring to these extraneous situations, are supposed to somehow transport these problems from the classroom to the 'real world'. But adding a bit of contextual 'dressing' to a classroom task cannot make a school problem real. In other words, the concept of 'real' problem solving fails to recognise the problem identified here, that is the difference between the *social context* of a problem and its *representation* as a task (even if the task representation refers to the social context).

In this chapter I consider three main topics:

1. theories of mental representation of concepts and knowledge;
2. theories of task representations;
3. theories of the social context of learning mathematics.

These topics raise two main issues. First, the relationship between 1. and 2. How do different task presentations effect the learning mathematics? Second, the problem of transfer of learning. How is knowledge learned in one social context transferred across to be applied in another. This is the subject of chapter 10.

### MENTAL AND TASK REPRESENTATION
In this section I shall consider the importance of both mental (internal) and task (external) representations, and the relationships that hold between the two.
### BRUNER
Building on the theories of the psychologist Jean Piaget and philosopher Charles S. Peirce, Bruner (1960, 1964, 1974) developed a theory of how knowledge is represented mentally. Bruner proposes that there are three modes in which

knowledge is represented in our minds. These modes of representation, which form the basis of all learning and understanding, are as follows:

## 1. Enactive
The enactive mode: this is understanding that can only be acted out, like knowing how to ride a bicycle, tie a shoe-lace, or use a pair of compasses. It is knowing 'how to' in a physical sense, rather than knowing 'that'. Enactive knowledge leads to knowledge in the iconic mode.

## 2. Iconic
The iconic mode: this is understanding through the use of mental pictures. For example, the concept of a circle is linked in many peoples' minds with a mental image of a circle. Much mathematical problem solving involves visualising situations. This is based in the iconic mode of mental representation. When developed, this leads on to knowledge represented or understood in the symbolic mode.

## 3. Symbolic
The symbolic mode: thinking with abstract symbols. Only in this mode can we understand formal mathematics like numbers, fractions, ratio and algebra.

Bruner argues that we first use these modes of representation and understanding in the order enactive – iconic – symbolic. However he argues that we continue to use all three modes, once we have developed them.

Bruner suggests that learning is well grounded if it passes through all three of these modes. Thus his theory has some useful consequences for the teaching of mathematics. He believes that learning should work through the three modes: enactive - iconic - symbolic in each topic, to provide a meaningful basis and development. Even when revising, we should re-establish the enactive and iconic bases of a mathematical topics before moving on to symbolic work. The table below shows the three modes applied to a number of mathematical topics.

*Table 9.1: Bruner's 3 modes of representation applied to mathematical topics*

| TOPIC | ENACTIVE | ICONIC | SYMBOLIC |
|---|---|---|---|
| *LOGO* | Moving own body or floor Turtle | Direct drive of Turtle on screen | Writing Logo procedures |
| *FRACTIONS* | Cutting objects into equal parts | Shading 'pie' diagrams | Numerical operations with fractions |
| *REFLECTION* | Cutting out, Using mirror | Reflecting pictures in mirror line | Computing reflections on Cartesian grid |
| *PROBABILITY* | Tossing coins, dice | Drawing tree diagrams | Calculating combined probabilities |
| *NUMBER* | Act of tallying | Tally marks | Use of numerals |
| *ANGLE* | Physical turning | Drawing of angle | Angle name and measures |

Of course this table only uses the symbolic mode to represent all three modes. Figure 9.1 below increases the range of modes to illustrate the teaching of two topics: fractions and reflection.

*Figure 9.1 Bruner's 3 modes of representation applied to 2 mathematical topics*

| BRUNER'S MODE | FRACTION | REFLECTION |
|---|---|---|
| ENACTIVE | The <u>actions of</u> cutting cake <br><br>sharing sweets 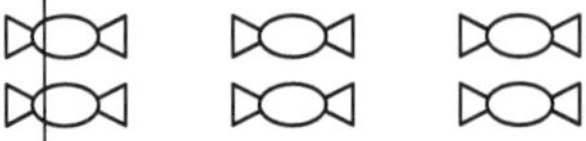 | <u>Actions</u> of reflecting in mirror 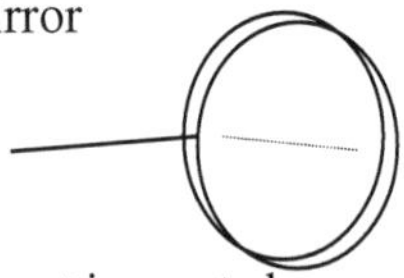<br><br>or cutting out shapes  |
| ICONIC | Diagrams 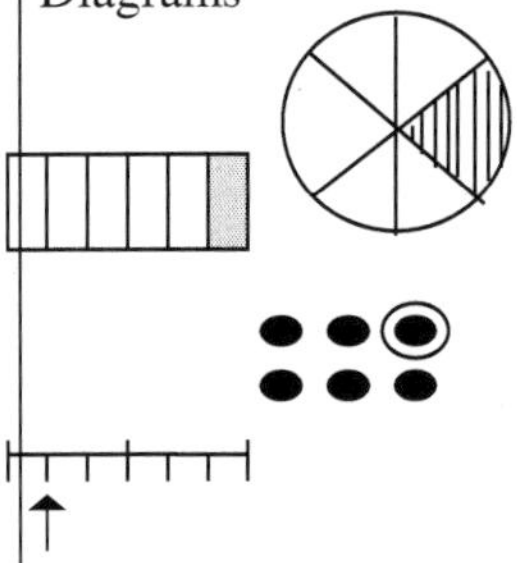<br><br>(mental representations of them) | Mathematical reflections in diagrams 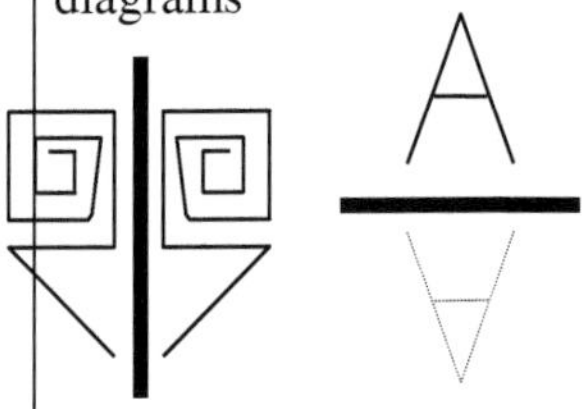<br><br>(visual understanding of reflection) |
| SYMBOLIC | Manipulating symbolic representations both mentally and on paper<br><br>$\frac{1}{6}$  $\frac{19}{31}$  $8 \div 17$<br><br>3:4  0.1  11%<br><br>$1\frac{7}{8}$  $\frac{1/2}{1/16}$ | Symbolic reflections<br><br>Reflect (6,2) in x axis (y=0) 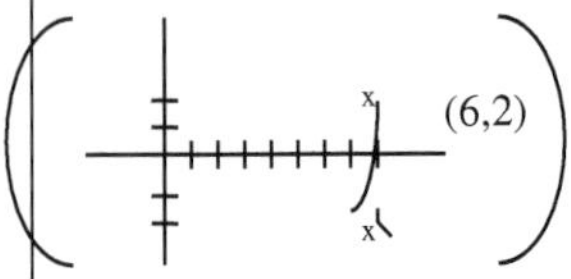<br><br>(6,2)<br><br>Reflect (1.5, 7.5) in y=x |

There is an analogy between Bruner's theory of representation and an ancient Chinese proverb: To hear is to forget, To see is to remember, To do is to understand.

According to the analogy, hearing corresponds to symbolic knowing, and is the last mode to develop and the least well grounded, hence the forgetting. Sight corresponds iconic knowing, which is better grounded, leading to remembering. Doing corresponds to enactive knowing, which is the best grounded, leading to a deep and lasting understanding.

Already there is a risk of confusion, because I have only illustrated how Bruner's three modes of representation can be applied to mathematical topics externally. That is, these are modes of task representation illustrated, not modes of mental representation. According to Bruner's theory, successful learning experience with these modes of external representation should lead to the development of corresponding (but also interconnected) mental representations in these modes.

## Z. P. DIENES

Another psychological researcher on the teaching of mathematics is Zoltan P. Dienes. He also proposes that that practical experiences are essential in mathematics learning. He is also a great believer in the value of games in the teaching of mathematics (Dienes 1960, 1964, 1970).

Dienes proposes two principles of mathematics learning concerning activity and the use of practical apparatus, which have been particularly influential:

### The perceptual variability principle.

Students should have several seemingly different practical experiences from which to build up the same mathematical concept or understanding. By experiencing the same mathematical concept in a number of different embodiments learners will be helped to extract the concept from the experiences.

Thus, for example, children should learn the concept of place value in number from the Dienes decimal blocks; from paper 100 squares, 10 strips and unit squares; from multilink tens and units; from match sticks (with bundles of 10 and 100); and from paper strips (with bundles of 10 and 100 clipped together). Each of these sets of apparatus provides an experience of grouping into tens and hundreds, and of decomposing into tens and units. As in this example, Dienes believes that students are more able to generalise from a variety rather than from a single experience.

### The mathematical variability principle.

Students should have practical experiences of similar but mathematically different apparatus. Such apparatus might have the same basic structure, but differ in terms of some number. An example of such apparatus is provided by abacuses in different number bases; or by Dienes' Multibase Arithmetic Blocks. He provides blocks to illustrate the number bases 10, 6, 5, 4, 3, and 2. In each base there are units (cm cubes), longs (strips as long as the base number), flats (flat pieces the size of the unit squared), and blocks (size of the base cubed). This apparatus shows that many of the properties of our number system, such as grouping and decomposition, depend on the structure, rather than on the base (ten, most commonly). The structural analogy can be seen by, e.g., comparing base 4 and base 10 Dienes blocks.

Dienes also proposed two further principles.

**The constructivity principle.**
Students need to construct their own concepts and understanding on the basis of their own activities and sense making. This is an early form of constructivism, as discussed in Chapters 6 and 7.

**The Dynamic Principle**
All concept learning with structured apparatus should go through three stages of activity:

1. Play stage, where children explore the apparatus freely – thus getting familiar with it.

2. Structured stage, where children are becoming aware of the structural properties embodied in the structured apparatus.

3. Formulation of concept followed by practice where children explicitly formulate the concepts embodied in the structured apparatus and then practice using and applying them.

Once again, Dienes' theory is about task representation, rather than mental representation. However his approach provides a good way, like Bruner's ideas, of providing secure learning experiences through which children can build up their own mental representations.

Dienes has created a great deal of very good mathematics apparatus for teaching number (Dienes 1960a), algebra (Dienes 1961), and many other topics including logic, measurement, ratio (Dienes 1966, Seaborne 1975) including mathematics learning games for young children (Holt and Dienes 1973).

Dienes, Bruner and indeed also Piaget have proposed that the learning sequence in mathematics should progress in the order practical - pictorial - symbolic. However, in this sequence an important factor is neglected. This is the role of language and discussion. Language plays a special role in understanding and the representation of meaning. Currently the ideas of Vygotsky (1978) about the role of language are growing in influence, and it is increasingly seen as not only an essential vehicle for learning, but also as playing an active part in the formation of concepts and understanding. Vygotsky's ideas about context and the goal directed nature of all activity, including learning, are also increasingly influential. Vygotsky's theory of learning is discussed in the next section, because it overtly brings in the social context of learning.

From the perspective of mental representations, so far the main emphasis in the discussion has been on the three modes distinguished by Bruner: the enactive, iconic and symbolic modes. Goldin (1987, 1992), in discussing the mental representations used in problem solving, has proposed a more extensive model of representation in mathematics including 5 clusters of factors:

- Imagistic systems (including spatial, auditory and tactile)
- Verbal or linguistic systems
- Formal symbolic notational systems
- A meta-cognitive system (for planning, monitoring and executive control)
- A system of affective representations (emotions, attitudes, beliefs, values)

He argues that not only do we need all of these modes on mental representation in mathematical thinking, but that they are all richly interconnected and linked, in the mind and thinking of the effective problem solver.

These 5 interlocking systems bring together the modes of representation considered here as well as those considered in the chapters on affect and problem solving. The model therefore includes virtually all of the factors discussed in this course.

There can be problems with the supposed links that are supposed to exist between different modes of task and mental representation. Hart (1989), Johnson (1989) and colleagues in the UK research project Children's Mathematical Frameworks found that children often fail to link practical work with the formal mathematics that is intended to develop from it. This often means that the formal mathematics - be it the decomposition method of subtraction, the formula for the volume of a cuboid, or the procedure for simplifying fractions - remains meaningless for the children, even though they have had the appropriate practical and concrete experiences. The problem is that children fail to build the conceptual bridge between concrete practical experience and a more formal understanding of mathematics. This is one problematic aspect of the transfer of learning.

The problem of transfer is as follows. Can what is learnt in one context be successfully transferred to another context? As long ago as 1912 Thorndike showed that learning traditional Euclidean Geometry did not necessarily transfer to improved reasoning skills - either logical or problem solving (Kilpatrick 1992).

Transfer of learning is of course central to problem solving, which was the topic treated in chapter 7. The whole point of acquiring general problem solving strategies is to be able to apply them to new problem situations and applications, that is, to transfer them to a new situation. Unless some transfer of learning was possible there would be no point in developing problem solving skills. However the issue is not straightforward as this and the following chapter show.

Perkins and Salomon (1989) reviewed the literature on transfer and concluded:

To the extent that transfer does take place, it is highly specific and must be cued, primed and guided; it seldom occurs spontaneously. ... When general Principles of reasoning are taught together with self-monitoring practices and potential applications in varied contexts, transfer often is obtained. (Quoted in Fey 1996: 20)

Research on the transfer of learning suggests that particular and tacit knowledge do not transfer well from the context of acquisition, whereas general and explicit knowledge are more susceptible to transfer (Kilpatrick 1992).

## THE SOCIAL CONTEXTS OF LEARNING MATHEMATICS

| *ANECDOTE* |
|---|
| Ann, aged 15, is a low attainer in mathematics who cannot do 2 or 3 digit written subtraction sums like 100-84 in standard column form. Ann however knew very well how to calculate and check her change for a packet of cigarettes costing 84p from £1.00 (prices as they were). |
| Why? Is this not the same mathematics? Is the difference that one sum is done at school, while one is meaningful 'street maths'? Or is it that different methods are involved (i.e., use of a written, formal algorithm versus informal mental/oral methods)? |

A growing body of research is suggesting that not only *what* students learn, but *where* they learn it is important. In other words, it is not only the *content* of learning that matters, but also the *social context* of that learning that counts. This also raises the problem of transfer discussed above. Can what is learnt in one social context be successfully transferred to and used and applied in another context?

A number of researchers in Brazil have been exploring the relationship between street and school mathematics, that is, the mathematics children and others perform on the street in their everyday economic activities, and in the classroom as part of their mathematics learning activities. Table 9.2 shows an example of some dialogue from such an investigation.

*Table 9.2 Sample dialogue from Carraher (1988)*

| **Carraher and her colleagues interviewed children vendor on the street in Brazil.** |
|---|
| Customer/examiner: How much is one coconut? |
| Child/vendor: Thirty-five. |
| Customer/examiner: I'd like three. How much is that? |
| Child/vendor: One hundred and five. |
| Customer/examiner: I think I'd like ten. How much is that? |

> Child/vendor: (Pause) Three will be 105; with three more, that will be 210. (Pause) I need four more. That is ... (pause) 315 ... I think it is 350.
> Customer/examiner: I'm going to give you a five hundred note. How much do I get back?
> Child/vendor: One hundred and fifty.
> When engaged in this type of interaction, children were quite accurate in their calculations: out of 63 problems presented in the streets, 98% were correctly solved. We then told the children we worked with mathematics teachers and wanted to see how they solved problems. Could we come back and ask them some questions? They agreed without hesitation. We saw the same children at most one week later and presented them with problems using the same numbers and operations but in a school-like manner. Two types of school-like exercises were presented: word problems and computation exercises. Children were correct 73% of the time in the word problems and 37% of the time in the computation exercises. The difference between everyday performance and performance on computation exercises was significant. These results convinced us of two things. First, street mathematics and school mathematics are not one and the same mathematics. Second, Brazilian schools do not acknowledge the existence of street mathematics, even if we all know of its existence through our everyday experiences. This appears to be an instance of what can be called the ideology of school mathematics: to ignore (or to treat as lesser mathematics) solutions which do not follow the school-prescribed ways. (Carraher 1988: 3-4)

Carraher and her colleagues decided to test if the mode of presentation was more important than the social context or location. So they gave the same arithmetical questions to a sample of children both oral, and in paper and pencil form. Table 9.3 shows the percentage of correct responses by operation and type of procedure used by the children.

*Table 9.3 Percentage of correct responses by operation and type of procedure used by the children (from Carraher 1988: 5)*

|  | MODE | |
| --- | --- | --- |
| **OPERATION** | Oral | Written |
| Addition | 75 | 68 |
| Subtraction | 62 | 17 |
| Multiplication | 80 | 43 |
| Division | 50 | 4 |

They found that for addition questions there was not a great deal of difference, but that when they put subtraction, multiplication and division tasks to the children, there performances were far superior in oral mode.

What can be concluded? In my view, that paper-and-pencil operations are different and harder quite often than oral or mental ones. So some of the difference between street mathematics and school mathematics can be attributed to the different mode of task representation and the associated methods. This concerns the task context in the representational sense. However this does not mean that the social context of the tasks does not have an impact on the students.

In case one is tempted to argue that children should be allowed to do their tasks in the mode at which they are most effective, we should remember that students need to master paper-and-pencil operations, not least because this is the mode of assessment used to certify mathematical knowledge.

Before returning to a discussion of the social context in learning mathematics, I want to explore these two modes of representation and thinking in mathematics further.

The difference between written, formal algorithms and mental, oral, and informal methods has been explored by Plunkett (1979). He contrast them as follows.

*Table 9.4 Plunkett's contrast between written algorithms and mental methods*

| **Standard Algorithms** Written | **Mental Methods.** |
|---|---|
| written | fleeting |
| standardised | variable |
| contracted | flexible |
| efficient | active |
| automatic | holistic |
| symbolic | constructive |
| general | not designed for recording |
| analytic | iconic |
| not easily internalised | early approximation to answer |
| cognitively passive? | limited |

Table 9.4 indicates the major differences between these two modes of mathematical thinking or activity. As argued above, students need to master both forms. Mathematics specialist teachers have always known that mental facility in mathematics is a vital skill. This is currently in vogue again with the UK National Numeracy Strategy.

## VYGOTSKY'S SOCIAL THEORY OF LEARNING[3]

Returning to the social context of learning, the most important social theory of learning is that of the Soviet Russian psychologist Lev Vygotsky. Vygotsky lived from 1896-1934, but his works first reached the West in translation in the 1960s and have only achieved widespread prominence in mathematics education research in the 1990s (Vygotsky 1978, 1979, 1986).

Vygotsky argues that thought and thinking depend on language that is acquired in discussion and conversation with others. Thus language actively used in discussion is a natural medium for teaching and learning. But language is always a part of a social context, a situation in which we are acting to achieve our goals, whatever they might be. In other words, it is the social context and situation that comes first, including the activities therein. Our response to these activities is secondary.

This has more revolutionary implications that it appears to have at first sight. For what it implies is that the stimulus for learning comes from outside the individual. Unlike the constructivists who see an individual's meaning construction as primary, Vygotskian theory sees this as secondary.

Building on his social theory of learning Vygotsky developed the theoretical idea of a learner's Zone of Proximal Development. This is that new area of thought and problem solving just beyond a learner's present unaided capacity, but which can be achieved which support from more capable others, whether teachers, peers, parents or others.

*Fig. 5.2: Vygotsky's Zone of Proximal Development*

**ZONE OF PROXIMAL DEVELOPMENT**

| Ability so far | Abilities within reach | Abilities currently beyond reach |
|---|---|---|
| what student can do unaided. | what student can do (can only do) with help of more capable other (teacher, peer or parent) | what student cannot (yet) do with help |

According to this view, abilities are something that are mastered by a learner though interaction and guidance by teacher, peer, parent or any other providing 'scaffolding' to support the development of competence. They are acquired from without, and then elaborated as they are internalised and appropriated, i.e., become the learner's own skills and knowledge.

---

[3] Followers of Vygotsky because of the primary importance he gives to language have adapted the Chinese proverb quoted above as follows: To hear is to forget, To see is to remember, To do is to understand, To discuss is to make my own. According to the new analogy, only when I can express my understanding in my own words do I truly come to own it.

As a mathematical example one may consider how children first learn to solve a particular set of routine problems, e.g., simple equations in algebra. This is often accomplished by following and then mimicking the actions of the teacher or another more capable other in solving similar tasks.

Vygotsky also argues that "human action, on both the social and individual planes, is mediated by tools and signs." (Wertsch 1991: 19). He distinguishes between spontaneous, intuitive concepts that the child first develops, and the subsequent higher level or scientific concepts, acquired through language. The latter leads to thinking which is qualitatively different from earlier thought. In order to facilitate and further their purposes, according to Vygotsky, human beings have created psychological tools, analogous to the use of physical tools in the world of work.

> Psychological tools are artificial formations. By their nature, they are social, not organic or individual. They are directed towards the mastery or control of behavioral processes – someone else's or one's own – just as technical means are directed toward the control of processes of nature. The following can serve as examples of psychological tools and their complex systems: language; various systems for counting; mnemonic techniques; algebraic symbol systems; works of art; writing; schemes, diagrams, maps, and mechanical drawings; all sorts of conventional signs; etc. (Vygotsky 1979: 137)

Thus Vygotsky explicitly recognises the function of mathematical symbol systems as tools which must be socially acquired and mastered (changing the learner in the process) in any acquisition of mathematical competence and knowledge. Thus a key stage in the learning of mathematics is the acquisition of competence in its written (and figural) linguistic forms and symbolism (in addition to mental and oral modes). For not only is collective mathematical knowledge recorded and transmitted largely by means of written text, but it also uses its own specialised and abstracted codes and symbolism. The gradual appropriation and mastery of this symbolism is an activity that takes the young learner of mathematics many years.

In the present discussion, the significance of Vygotsky's theory of learning is that it links the content that is learned with the social context in which it is learned, including relationships with other persons, and all of the expectations, emotions, roles involved.

Some social theorists argue that different social contexts trigger different aspects of yourself, and that different skills, abilities and knowledge are available to you in those contexts. This brings us back to the anecdote at the start of this chapter. Ann cannot do 2 digit written subtraction sums, but can calculate and check her change accurately. In school her role is that of an incompetent low attainer, who has experienced failure repeatedly. On the street she is a cheeky, socially capable young woman who smokes and acts like a functioning grown-up. Do the differing roles, expectations, and emotions bring out two different personalities, one of which is capable and the other incompetent? (Of course the different modes of calculation, as in the Brazilian street mathematics, are also relevant here.)

## SITUATED COGNITION

A theoretical approach to the problem of social context and transfer of learning which builds on the work of Vygotsky is that of 'situated cognition'. This has been developed by Brown *et al.* (1989), Chaiklin and Lave (1993), Lave (1988), Lave and Wenger (1991), Rogoff and Lave, J (1984) and others. They claim that what we learn is relativised to the social context of acquisition. That is the social context, or 'social frame', through associations triggers or otherwise elicits the skills and knowledge from long term memory. They claim that knowledge is stored separately according to the context of learning. "learning and cognition ... are fundamentally situated" (Brown *et al.* 1989, quoted in Fey 1996: 20). A consequence is that transfer will not occur unless links or bridges between different contextual domains of mathematical knowledge are constructed or made by some creative act of mentally linking what was learnt separately (Lave 1988, Lave and Wenger 1991).

Thus some parts of learners' mathematical knowledge relate to shared social activities, namely the contexts of acquisition and use, and are not easily detached from them or transferred. In particular, such aspects of mathematical knowledge are often elicited in their contexts of origination as part of an automatic component of recognition and engagement with the specific social situations to which they relate.

Partial confirmation of this approach can be found by reflecting on the divisions between school subjects. Even though they are studied in the same institution, students usually do not connect what they learn in mathematics, science, technology, let alone non-scientific subjects, even if the same mathematical knowledge and skill (from our perspective) are involved.

Although not based in the same theoretical perspective, Hart (1989), Johnson (1989) and colleagues in the project 'Children's Mathematical Frameworks', which found that children often fail to link practical work with the formal mathematics that is intended to develop from it. This often means that the formal mathematics - be it the decomposition method of subtraction, the formula for the volume of a cuboid, or the procedure for simplifying fractions - remains meaningless for the children, even though they have had the appropriate practical and concrete experiences. The problem is that children fail to build the conceptual bridge between concrete practical experience and a more formal understanding of mathematics.

Further work on situated mathematical activity in non-developed countries has been carried out by Saxe (1991). One of his central innovations is to remark the centrality of goals in out-of-school mathematics. He looked in particular at the understandings of Brazilian street candy sellers (and also earlier at Papua and New Guinea traders). He suggests that such goals change and emerge dependent on 4 sets of factors. knowledge

SOURCES OF EMERGING GOALS
1. Social interactions with various others
2. Conventions, money denominations, numerical systems
3. Prior understandings, and the new understandings that mastering notational forms bring
4. The structures of the activities they engage in

There is no doubt that goals represent an important if understudied aspect of mathematics and learning.

Simon Goodchild (1994, 1997) studied the goals of learners using SMP 11-16 and comparing these with those of the class teacher. He finds that often learners are just 'doing' mathematics., with little idea of the intended goals.

# CHAPTER 10

## MATHEMATICAL KNOWLEDGE AND CONTEXT

In this chapter I explore the question of what light a modern epistemological perspective can throw on the problems of situated cognition and learning in a social context. The analyses and accounts I offer are tentative and evolving, and intended as provocations and reflections indicative of evolving and unfinished thoughts. I express some of my concerns about the problem of the transfer of skills and knowledge from one context to another, but I do not claim to offer a state of the art survey of situated cognition.

The past half century has seen important shifts in conceptions of knowledge including the recognition of the explicit-tacit knowledge distinction (Ryle 1949, Polanyi 1958, Wittgenstein 1953). Explicit mathematical knowledge includes propositions with warrants, such as Pythagoras's theorem. Knowledge of proofs, problems and definitions can also be explicit, but most personal knowledge in mathematics is, I want to claim, tacit. Tacit mathematical knowledge includes methods, approaches, symbolic operations, strategies and procedures which are often applicable to new problems, but are used differently in different situations. For example, the column addition algorithm, proof by mathematical induction, and specific problem solving strategies such as holding one variable constant and examining the resultant pattern of values, are all procedures or methods which, I wish to claim, are largely known tacitly. Hence while the applications of these procedures and strategies are explicit, the more general knowledge underpinning them normally is not.

However the notion of application is a problematic part of the relationship between knowledge and context. Thus an important question concerns the extent to which tacit knowledge is applicable to new situations and what applying it to a new situation might mean. How widely are mathematical procedures and strategies applicable, and when are such applications new? More generally, what features are involved which individuate a context and which distinguish two contexts (i.e., one is 'new' relative to the other) or render two contexts or situations equivalent? (i.e., they are regarded as mathematical the 'same'). I cannot answer these questions, but I wish to signal their import.

First of all I want to make the distinction between explicit and tacit knowledge. Traditionally philosophy and epistemology have focused on explicit knowledge and talked about the warrant for that knowledge. I think we need to accommodate that type of knowledge but also make a space for tacit knowledge. There is a strong precedent for this. Partial parallels exist between a number of dichotomous classifications of knowledge. Thus corresponding to explicit knowledge there is propositional knowledge, which is commonly distinguished from practical knowledge, skills, dispositions. There is also Ryle's (1949) 'knowing that' versus 'knowing how'; there is Polanyi's (1958) and Kuhn's (1970) explicit knowledge versus 'tacit' or personal knowledge; there is Wittgenstein's (1953) explicitly stated knowledge versus the knowledge implicit in 'language games' and 'forms of life'. In our own field of mathematics education there is Skemp's (1976) and Mellin-Olsen's (1981) relational understanding versus instrumental understanding; and there is Hiebert *et al.*'s (1988) conceptual knowledge versus procedural knowledge. In each of these dichotomies the first of the pair of terms corresponds to explicit knowledge while the second term corresponds to tacit knowledge or 'know how'.

How can this be further elaborated in a way fruitful for the understanding of mathematical knowledge? Philip Kitcher (1984) proposes a model of mathematical knowledge drawing upon Kuhn's (1970) analysis of scientific knowledge. Kitcher calls it a model of mathematical practice, but it does not correspond with mathematical practice in any social sense. So I want to adopt it as model of mathematical knowledge, even though this goes beyond his intentions, which I have done more extensively in Ernest (1997). Table 1 shows the model, and includes Kitcher's components as the first five in the list. I have added two further components at the end of the list because these seem to be important items that are missing from the list, or are only present implicitly. Kitcher does not claim his list to be complete, so it is legitimate to add additional ones if, as I believe, they are needed.

*Table 10.1: A Model of Mathematical Knowledge (Based on Kitcher 1984)*

| Mathematics Knowledge Component | Explicit or Tacit |
|---|---|
| Accepted propositions and statements | Mainly Explicit |
| Accepted reasonings and proofs | Mainly Explicit |
| Problems and questions | Mainly Explicit |
| Language and symbolism | Mainly Tacit |
| Meta-mathematical views: proof & definition standards, scope and structure of mathematics | Mainly Tacit |
| Methods, procedures, techniques, strategies | Mainly Tacit |
| Aesthetics and values | Mainly Tacit |

First of all Kitcher includes accepted propositions and statements as mathematical knowledge, and those are mainly explicit. Secondly Kitcher includes accepted reasoning and proofs. Typically proofs are rigorous warrants in mathematics and are fully explicit. Including less formal reasonings opens up the range of items referred to. Accepted reasonings as discursive entities are mainly, if not totally, explicit. Problems and questions are circulated in discussion and between mathematicians and once again, these are mainly explicit. Kitcher includes two further areas. One is meta mathematical views, including views of proof and definition and views of the scope and structure of mathematics as a whole. This type of overview and general views are mainly tacit elements of mathematical knowledge. They are tacit in the sense that mathematicians get a sense of them and build them up incidentally through experience and are not and probably cannot be fully taught explicitly. These elements are usually acquired from experience and are tacit. Kitcher also includes language and symbolism as a further component and these are also largely tacit. Some aspects of knowledge of the language and symbolism of mathematics are known explicitly, but much of their use is tacit, and there are irreducibly tacit elements to this knowledge.

In addition Table 10.1 includes two further categories not proposed by Kitcher but which are important in discussions of mathematics education. First of all, the methods, procedures, techniques and strategies are important in the context of school mathematics and also in applications of mathematics, but seem to be omitted by Kitcher. Many mathematical methods do not fit under the other categories, and these are mainly tacit elements of knowledge. Maybe some elements of this category are explicit, but like an iceberg, supporting the explicit part is a large body of further knowledge that is tacit. Finally, the second additional category is that of aesthetics and values. In part this is similar to the metamathematical views, but it seems worth singling out as another element, since the values aspects of metamathematical views are not mentioned by Kitcher. Although explicit statements about the aesthetics and beauty of mathematics have been made by mathematicians such as Hardy, most person's positions and feelings about this are tacit, tied into personal beliefs and views which are at best only partly articulated.

The model of mathematical knowledge shown in Table 10.1 is evidently a broadening and an extension of the traditional view of knowledge as primarily explicit. The wider nature of the elements it includes means that it is more able to describe the practices of mathematicians and the processes of learning mathematics, since these include tacit elements. It proposes that mathematical knowledge includes a tacit and concrete dimension, made up of knowledge of instances and exemplars; of problems, situations, calculations, arguments, proofs, applications, and so on. This part of knowledge in mathematics, and I also believe in school mathematics, comes from the experience of working with mathematics, and a lot of it is built up tacitly as 'know how' rather than as explicit knowledge. It has already been pointed out by Schoenfeld (1985) and others that mathematical problem solving depends on concrete knowledge of instances, and past problem solutions. Thus there is a mathematical craft knowledge based on concrete particulars and instances which is vital in mathematics and learning mathematics, and much of this is known tacitly, or as knowledge of cases, examples, etc. This new emphasis on the tacit and particular is contrary to the widely held perspective that emphasises the import of abstract and general knowledge at the expense of tacit, concrete and specific knowledge.

It is worth remarking on the parallel that can be made here with the contrast between the scientific and interpretative research paradigms in educational research methodology. One of the features of the interpretative research paradigm is that it valorises concrete particulars and personal knowledge. Thus there is a parallel between the revaluing of tacit and particular knowledge in mathematics, proposed above, and the growing acceptance of the interpretative paradigm in educational research. This is not to denigrate scientific paradigm research nor the value of the generality to which it aspires. It is rather to note the growing value attached to tacit and personal knowledge, and to case studies and particulars in research. This growing new emphasis complements the explicit and general knowledge associated with the scientific research paradigm.

## *THE SOCIAL CONTEXT OF KNOWLEDGE AND TRANSFER*

Knowledge acquisition skills are socially acquired and knowledge is usually learned in social contexts. I do not think this assertion is controversial, provided that the concept of social context is interpreted widely enough (Ernest 1998a). I know a mathematician who studied and learnt university mathematics by reading Bourbaki's Elements in French on his own as a precocious teenager. But I would argue even this solitary learning activity was socially based, because he mastered the language and underlying knowledge socially, i.e., in conversation with others, and then exercised them on his own.

Raising the issue of the social context of knowledge involves a way of viewing knowledge that is alien to the received view in epistemology. For knowledge understood as warranted true belief has been viewed as independent of its social context, or even of the context of its acquisition. The context in which someone comes to know has been viewed as inessential to the status of the knowledge, which is secured by the logical context of its justification, not the contingent social or personal context of its genesis (Popper 1959). Even scientific knowledge which involves knowledge of the world as empirical generalisations transcends the instances of its testing, unless it fails such tests and thus is not knowledge anyway.

Such views represent one way of conceptualising knowledge, and indeed constituted a dominant way. However, they embody a Cartesian or post-Cartesian dualism, one that separates the realm of mind and knowledge, from that of bodies and the tangible world. Thus although logical positivism and logical empiricism reject a literal Cartesian dualism, they reintroduce what I would call a post-Cartesian dualism by distinguishing as ontologically distinct the realm of the analytic, *a priori*, and logic, i.e., that of reason, from that of the empirical, the *a posteriori*, i.e. from the mundane world of brute contingent fact. Clearly such views have implications for how the relationship between knowledge and social context is conceptualised and how knowledge application and transfer are viewed. In particular, these dualisms view the link between knowledge and context as at best weak, and thus the transfer and application of knowledge as relatively simple.

As my account implies, there are differing perspectives on knowledge application and use and I want to contrast the above view that knowledge is transferable with the view that knowledge is situated. In contrasting these views I do not wish to oppose them as a fixed dichotomy but instead recognise that many perspectives are possible and seen their contrast in simplest form rather as two poles of a continuum. The view that knowledge is transferable sees knowledge as transportable, i.e., it travels easily with the possessor, the knower, and is thus effortlessly transferable to a new context. As I have indicated this is a view that is quite widespread, being based on traditional and perhaps unexamined epistemological and ontological assumptions, and is often found in the pronouncements and edicts of bureaucrats and policy-makers. For example, the current vogue for the identification of personal transferable skills in higher education is sometimes based on a conceptualisation of the issue of the inter-contextual transfer as unproblematic. (Below I offer another more defensible, in my view, interpretation of personal transferable skills in terms of problem solving capabilities.) Such views contrast with the notion that knowledge is situated, and that knowledge remains linked to the context of acquisition, representing the other pole of the continuum described above.

The traditional epistemological and ontological views underpinning the notion that knowledge is easily transferable are usually associated with a further notion that there is a unique self or cognising subject separate from knowledge. Several of the other papers in this volume refer to subjective knowledge and subject-object relations, and it may be that there is often a presupposition that these relations are to do with separation. Namely, that the knower is quite independent of any knowledge, and is an agent who can grasp, have, or transport knowledge as if it is an independent entity. Indeed the metaphor of material possession of knowledge (to grasp, get hold of, or carry) presupposes this separation and the very mutable and impermanent relationship between knower and knowledge as a commodity. In contrast, a situated view of knowledge is often associated with a further, different view, and to put it in a neutral way that is consistent with different conceptualisations of the situatedness of knowledge, i.e., its indissoluble link with a social context, that there are multiple facets to the self, and that the knower and the known are related and context dependent. There are a number of different ways of elaborating these issues from different perspectives, as these contrasts indicate. Below six different perspectives on the transferability of knowledge, the relation between knowledge and social context, and the associated concept of self are distinguished.

First of all, there is the perspective that knowledge is universally applicable, based on the assumption that it is abstract and unrelated to context, and typically has the form of explicit propositions or laws expressing relationships. Consequently general knowledge can be applied in contexts by instantiation, through which the specific variables of a concrete situation are interrelated in a structural application of the knowledge, analogous to the process of substituting a particular set of parameter values into equations and formulas. In this way knowledge, like a scientific theory, is fully applicable in new contexts. Furthermore, from this perspective, the knowing self is entirely disjoint from both the knowledge and the context of application, and therefore, for all intents and purposes, can be factored out. This perspective corresponds to what was described above as the post-Cartesian dualism of logical empiricism. It separates the logical realm of abstracted knowledge from the concrete realms of the mundane and subjective (Popper 1979). This perspective does not acknowledge that there are significant differences between contexts, for contexts can only be used to test and possibly falsify knowledge, and not to generate new knowledge (at least not in their epistemological function). Thus the concept of situated knowledge is incoherent from this perspective.

Although this a self-consistent and defensible position, it does not accommodate the broader view of knowledge summarised in Table 10.1, and hence does not admit as legitimate, let alone address, the problems of transfer discussed here. In particular, since knowledge is universally applicable, it makes no sense to say it is transferable or transportable, because there is no origin or location from which to transport/transfer it, since it is located in logical space - not in physical or social space. This perspective corresponds to views of knowledge and learning of traditional epistemology. Knowledge is knowledge because of its justification, and learning is established by the evidence of assessment and both, once validated, are context independent.

Second, there is the modelling perspective which links abstract academic knowledge with the concrete knowledge of specific application contexts dialogically. According to this view, explicit knowledge is fully portable and can be applied in any situation through modelling. Knowledge of contexts of application can also be imported into the context of learning mathematics as a basis for concept development and problem solving. Thus there is a two way traffic between the academic and applications contexts. According to this view, the self is detached from knowledge although new personal knowledge can be induced from immersion in concrete situations, i.e., the context of application.

Third, there is the view that knowledge exists in both explicit and tacit forms, and knowledge which is explicit and abstract is transferable. According to this view tacit knowledge is embedded in certain task specific capabilities. To make knowledge transferable it must first be disembedded from specific tasks or contexts, and transformed into explicit and abstract form. Once knowledge is expressed this way it becomes transferable and transportable, similar to the first perspective. The disembedded and abstracted knowledge is applied and hence re-embedded in a new task context, hence achieving transfer. Underlying this view is the assumption that self and knowledge are separate, but it is understood that tacit knowledge develops as a consequence of experience with specific sets of problems or situations. This perspective acknowledges that problems of transfer exist, but conceptualises transfer in cognitivist terms. Thus transfer is seen as concerning different sets of tasks which vary according to cognitive demand, mode of representation, and perhaps other variables. Transfer of learning thus concerns applying skills and knowledge learned for one set of tasks to another. This perspective thus corresponds to cognitivist views of knowledge and learning.

Fourth, there is the problem solving perspective of transfer that sees a person's higher order problem solving skills as transferable. This perspective acknowledges that both explicit and tacit knowledge exist but it emphasises the individual ownership of knowledge, especially with regard to the tacit knowledge of problem solving. According to this view, the bulk of a person's tacit knowledge including strategic knowledge cannot be made explicit. Instead this knowledge can only travel with a person and is made relevant to a new context of application by the person's immersion in the new context and the cumulative experience of working there. In applying their knowledge in a new context individuals are having fresh learning experiences as well as relating existing knowledge to the new tasks. Personal knowledge is developed through this experience becoming an additional personal resource and knowledge base, extending the person's capabilities without affecting the nature of selfhood. In anticipation of this development, cultural resources from the context of application can be imported into a learning context to prepare the learner for problem solving in the context of application. This perspective conceptualises the problem of transfer primarily in terms of personal cognitive capacities which can be further developed in different domains of knowledge application and problem solving. The difference between different domains of knowledge application is not that they are distinct social contexts, but that they are sites combining the application and acquisition of task specific knowledge and capacities with affective factors, i.e., goal orientations. This perspective corresponds to a problem solving view of mathematics, as fitting well with constructivist views of learning.

The fifth perspective views knowledge as partly situated within the social context of its generation and use. In consequence, some know-how or personal capacities cannot be divorced from their context of origin, but are elicited there by the combination of cues and the personal demands that the context makes. According to this perspective some elements of knowledge can be recontextualised and further developed, as new situated knowledge is created within a further social context, if a knowledgeable person moves across and works in the new context. According to this view the self and knowledge are interrelated. The self has multiple but connected facets each of which is elicited with its associated knowledge and capabilities in the appropriate social context. Thus the social context acts an enabler, providing an appropriate set of personal roles, positionings, interpersonal relationships, expectations, tools, resources, and characteristic activities and tasks, which enables a person to activate a range of capacities and skilled performances. However this complex nexus is socially situated and acts as a whole, and it is inappropriate to think of elements of it being rationally selected and reassembled in another context. Thus the social context is, to a greater or lesser extent, an indivisible whole.

This perspective adopts a situated cognition view of knowledge, learning and transfer. The problem of transfer is not conceptualised as being simply the application of skills learned in one problem set to another. Instead it is conceptualised in social terms: how can the capacities and knowledge and intellectual resources and tools developed for use in one social context be redeveloped, extended and redeployed in another? This perspective approximates to that of situated cognitionists, post-structuralists, social constructionists, and related theorists.

Sixth, there is the more extreme view that knowledge is completely situated and cannot be divorced from its context at all. According to this perspective individuals must be apprenticed within the context of a social activity to master its situated knowledge, and there is little of significance that can be transferred in or out. Discrete segments of the self are developed in different contexts and their resources and dimensions are enabled only in those contexts. Transfer in anything but a trivial sense is eliminated.

This sixth perspective in pure form probably has few adherents, although it is invoked as a 'straw person' target for attacks on situated cognition or post-modernist perspectives. Virtually all scholars would acknowledge that there are some core elements of selfhood that are transported to any context in which the person is engaged. Otherwise the person could have no knowledge of contexts other that in which s/he was engaged at the moment, and questions would need to be raised about access to personal linguistic knowledge and other resources in multiple contexts.

The six different theoretical perspectives described above are summarised in Table 10.2. below. In the table it should be noted that across the range of different perspectives presented the concept of transfer of training, learning or knowledge has distinct meanings. In the first and last it has little meaning because knowledge is not associated with context, in the first case, and knowledge cannot be dissociated from context and transferred, in the last. So I shall disregard these extreme cases, which merely serve as markers at the extremes of the range of

possible positions on transfer. The four remaining meanings of transfer are those of the applied mathematicians, cognitionists, problem solvers or constructivists, and situated cognitionists or social theorists.

*Table 10.2: Different Perspectives on the Transferability of Knowledge*

| Perspective | View of Knowledge | View of Transfer | View of Self |
|---|---|---|---|
| 1. Knowledge is universally applicable | Knowledge is abstract and unrelated to context | In strict terms, there is no transfer. General knowledge has the specific variables of the concrete situation inserted for full applicability | Self entirely disjoint from knowledge and context |
| 2. Modelling links abstract with concrete | Abstract academic knowledge and concrete knowledge of specific application dialogically linked through modelling | Knowledge fully portable and can be applied in any situation through modelling | Self detached from knowledge although new personal knowledge induced from concrete situation |
| 3. Explicit knowledge is transferable | Knowledge exists in both explicit/abstract and tacit forms. | Explicit knowledge is transferable. Tacit knowledge must be made explicit and abstract before it becomes transferable | Self and knowledge separate, but tacit knowledge developed in contextual experiences |
| 4. Personal problem solving skills transferable | Persons have both abstract explicit knowledge and tacit knowledge, including strategic problem solving knowledge | Tacit knowledge cannot all be made explicit, but is transported with person and made relevant by experience via immersion in context of application | Personal knowledge developed through experience becomes additional personal resource but does not affect core of self |
| 5. Knowledge partly situated | Some knowledge cannot be divorced from context | Some elements of explicit knowledge can be recontextualised and further developed as new situated knowledge is created | Self has multiple but connected facets each of which is elicited with associated knowledge in its context |
| 6. Knowledge fully situated | Knowledge cannot be divorced from its context | Individuals must be apprenticed in new context to master situated knowledge | Discrete segments of self developed in different contexts |

The four middle perspectives interpret transfer as:

1. Transfer of learning is application: applying general knowledge to specific concrete situations via modelling
2. Transfer of learning from one set or type of tasks to another – the transport of disembedded knowledge
3. Transfer of learning from one problem situation to another through transport of personal transferable skills (with a person)
4. Transfer of learning from one social context to another through the development of new capacities and facets of self

Just as the concept of transfer needs to be analysed and as here shown to have multiple meanings, the concepts of context and social context need clarification. In particular, what individuates a context, and the criteria for the equivalence of two different contexts is important, as I said above. For until it is determined when two contexts differ or count as the same, it cannot be stated with precision whether successful accomplishment of parallel tasks in the two counts as transfer of learning or simply represents the exercise of the same or analogous skills. However, in my view, the criteria for equivalence and individuation of contexts cannot be uniquely specified, for these will vary according to the perspective adopted including the interpretation of the concept of transfer.

My discussion thus far has been couched in general terms, and has barely touched on mathematics-specific features. Below I consider some of the mathematics-specific features that arise in a consideration of contextual and pedagogical dimensions of transfer from an educational perspective.

## Applications Perspective

According to the applications perspective, modelling builds a link with the applications context. The claim of this perspective is that the problem of transfer is thus overcome. The 'real world' context of the application and the academic or school mathematics contexts are in a dialogical relation which builds permanent bridges between them, connecting both. First of all, representations from the context of application provide the basis for generating concepts methods and problems in the Academic or school mathematics context, via abstraction and generalisation. So there is a flow from the context of application to that of schooling. Second, there is a flow in the other direction. abstract mathematical knowledge, concepts, skills and models in the school context are used in applications, and verified in the context of application. Ultimately, when the knowledgeable user of mathematics is immersed in applications, the academic/applications context difference becomes irrelevant. For mathematical models can be formulated in either. The important difference becomes that between the abstract level of models and the concrete 'real world' level of empirical problems, solutions and data. The applications perspective on transfer and inter-contextual relations is illustrated in Figure 10.1:

*Figure 10.1: The applications perspective on transfer and inter-contextual relations in maths*

| 'Real world' context | Generation<br>→<br>abstraction | Academic/school maths context |
|---|---|---|
| Concrete and applied maths problems and applications | Application<br>←<br>verification | abstract mathematical knowledge concepts and skills |

So this is one way of conceptualising transfer and inter-contextual relations; one that does not see transfer as problematic. It reduces the need for transfer by quilting together the contexts.

## Cognitivist Perspective

According to the cognitivist perspective explicit mathematical knowledge learned in the school maths context is transferable to external uses in the 'real world' contexts of numeracy and mathematics. Explicitly learned school mathematics, including symbol systems and computational algorithms, as disembedded knowledge, is applicable to mathematically-susceptible tasks originating in domestic, popular, work and other external contexts. This facility depends on the ability to identify and then work mathematical tasks located in these external situations. This facility is developed by the importation of characteristic elements of these 'real world' contexts. These are external task representations together with some incidental features associated with them to help future identification. The cognitivist perspective on transfer and inter-contextual relations for mathematics is illustrated in Figure 10.2 below.

*Figure 10.2: The cognitivist perspective on transfer and inter-contextual relations*

|  |  | School maths context | Context of work and employment | |  |
|---|---|---|---|---|---|
|  |  |  | import ↓ | |  |
|  |  | School maths context | Work type application tasks |  | 'Real world' context |
| Domestic / popular context | import → | Domestic / popular school tasks | Standard mathematical tasks | transfer ⇨ | numeracy and maths |
|  |  | School maths context | Ethnomathematical and cultural tasks |  | uses and applications |
|  |  |  | import ↑ | |  |
|  |  |  | ethnomathematical and cultural context | |  |

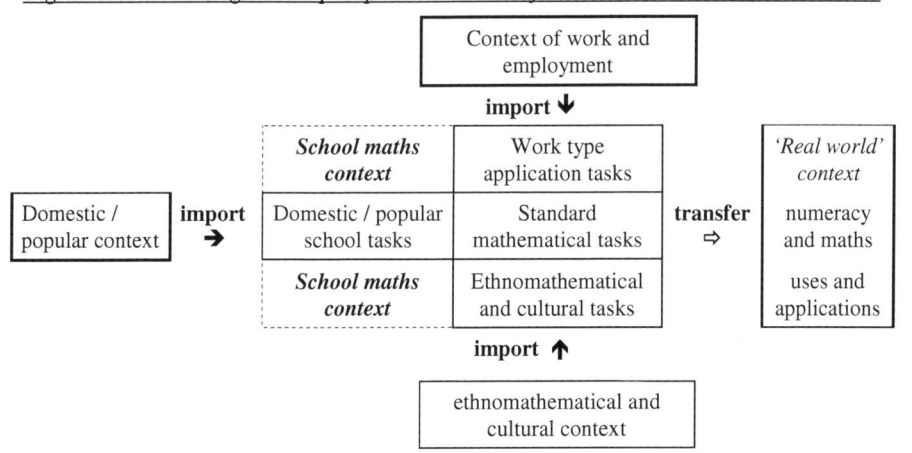

Figure 10.2 illustrates the transfer of explicit mathematical knowledge from the school context to external 'real world' contexts, which this perspective understands to be unproblematic. This is the implicit model presupposed as underlying progressive mathematics education. The two main contexts in the diagram are the academic school mathematics context and the so-called real world context where numeracy and mathematics are applied. In this model there is the naive assumption is that there is such an entity as 'knowledge' - and knowledge related skills are transferred into the real world context. The figure also illustrates three external contexts which are used as sources of task representations imported into the school

mathematics context. These are the world of work and employment, with imported tasks such as calculating tax deductions from wages; the domestic / popular context of hobbies, shopping, etc, with tasks such as modifying cooking recipes, or calculating discounts in sales; and the ethnomathematical and cultural context, with tasks such as drawing Rangoli patterns or Islamic-style tiling patterns. Figure 10.2 distinguishes the actual contexts which provide the inspiration for such tasks from the sub-domain of the school mathematics context containing these task representations. It is widely regarded as good practice within this perspective to introduce a degree of authenticity into tasks by importing detailed representations to provide partial resemblance to the original context-embedded tasks. Thus realistic looking wage slips, advertising brochures illustrating sales goods with prices and discounts, and photographs of Islamic-style tiling patterns in the Alhambra would serve this purpose.

We import representations and elements from the domestic/popular context, and they get transformed, recontextualised or become the inspiration for writing domestic popular tasks for the classroom. Part of the rationale is that this importation provides a conceptual foundation on which mathematical learning is to build, through tapping into meaningful out of school experiences and knowledge. It is also intended that these tasks will be motivational, because out of school activities are purposive and goal directed. The same holds for work context-related tasks, because these are meant to be useful for students in their working lives. Such tasks are also intended to directly useful, by facilitating transfer into the adult workplace. Thirdly, the importing of elements from the ethnomathematical and cultural contexts of mathematics, particularly informal mathematical reasoning and patterns used in non-European countries, is likewise meant to provide an authenticity to academic school mathematics, and also to facilitate transfer. It is also meant to have socio-political implications, by raising awareness and valorising the products of non-European cultures (Powell and Frankenstein 1997).

Each of these three types of import involves redundant elements of representation, in terms of the underlying school mathematical task. But this redundancy serves to make the tasks appear 'authentic', i.e., as if they were being attempted in the external contexts. As many have remarked, such 'contextualisation' is a form of decoration and can never close the gap between the school activities and authentic context-bound tasks. However it can provide practice in the skills of identifying embedded tasks before applying mathematical symbolisation and procedures. Lastly, Figure 10.2 shows the central sub-domain of the school mathematics context coining standard mathematical tasks. These are 'undecorated' with is no direct reference to any extra- mathematical context, and comprise the straight forward application of mathematical procedures to symbolically encoded, i.e., routine, mathematical tasks.

## Problem Solving Perspective

The problem solving perspective views the most important knowledge for transfer as tacit personal knowledge, namely problem solving strategies and heuristics. This knowledge is acquired primarily from solving non-routine problems in the school context, plus from seeing teachers and others showing solution methods for particular problems. Another possible source of this knowledge might be explicit instruction in mathematical heuristics, although the jury is still out on whether this adds anything worthwhile to knowledge. One of the complexities of problem-solving knowledge is that it can only be learnt from a finite number of exemplars (plus the other possible sources mentioned above) and then somehow it becomes transferable to an unlimited number of examples. Clearly students can only have a finite experience of exemplars from which patterns of heuristics are generalised and induced, providing the basis for transfer. Success in this process seems to depend on the transferable meta-skill of being able to learn from and generalise the knowledge from the instances in the first place. Perhaps this skill corresponds to what Bateson (1972) calls deutero-learning?

The key feature of the problem solving perspective is that the most significant inter-contextually transferable skills are the personally acquired, personally transportable heuristics and higher-level skills. The problem solving perspective on transfer and relations in mathematics is illustrated in Figure 10.3.

*Figure 10.3: Problem solving perspective on transfer and inter-contextual relations*

| Academic/school maths context | Transfer | 'Real world' contexts of work |
|---|---|---|
| Acquired problem solving strategies and personal transferable skills | ➔ with person | Use of skills in practical and work-related problems and situations |

An important current issue concerning the transfer of skills which fits under this perspective is that of 'personal transferable skills'. Currently in further and higher education there is an emphasis in this area and curriculum developers are required to specify the personal transferable skills addressed in any teaching module irrespective of discipline, content or aims. For example, at Exeter University six clusters of personal transferable skills have been identified (self-management, learning skills, communication, teamwork, problem-solving, data-handling skills) and we are required to list all the personal transferable skills in our courses. It is not clear what is the theoretical basis for identifying these skills. The cognitivist view would be that if we disembed these skills and make them explicit they can be transferred to and reapplied in new 'real world' contexts. But these skills are primarily strategic, like problem solving heuristics, and such higher level skills by their very nature cannot be made fully explicit. For once strategies are fully explicit and determinate they become algorithms, and lose some of their heuristic quality. Thus it seems more appropriate to consider them with regard to the present perspective, since it concerns the personally acquired and developed strategic skills of problem solving etc, which it regards as transportable with the acquirer.

## Situated Cognition Perspective

The situated cognition perspective is that mathematical knowledge is partly situated and some of it cannot be divorced from its context of origin and deployment. This is what I understand situated cognition, legitimate peripheral participation or the Lavian or social anthropological view to be about (Lave and Wenger 1991). Thus my fourth picture of transfer and inter-contextual relations depicts a number of separate contexts. We have the school mathematics context and some other contexts of which samples are shown in Figure 10.4, including the domestic and popular context(s) of numeracy and maths use, the industrial and work context of maths applications, the academic university maths context. I have distinguished these because they involve different aims, roles, functions and practices, and there is discussion of the problem of transfer from one of these to another. The situated cognition perspective on transfer and inter-contextual relations is illustrated in Figure 10.4:

*Figure 10.4: Situated cognition perspective on transfer & inter-contextual relations*

| domestic/ popular context | Transfer? → | SCHOOL MATHS CONTEXT | ← Transfer? | Academic/ University maths context |
|---|---|---|---|---|

↑ Transfer?

| Industrial and work contexts of maths applications |
|---|

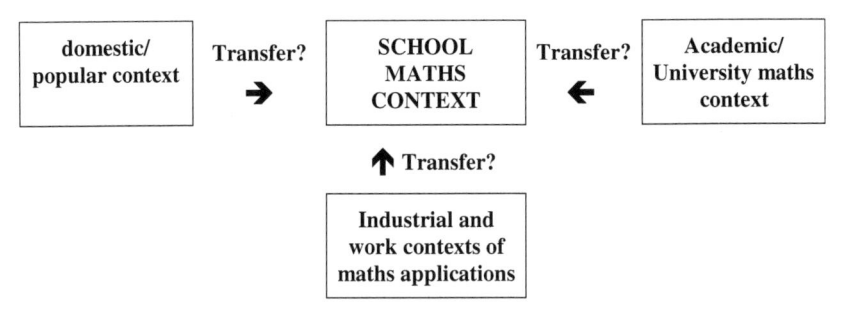

Figure 10.4 depicts the four contexts shown as discrete social practices, which emphasises the problem of explaining how knowledge and skill are transferable from one context to another. If they are all separate what are the relationships between the discrete social practices? This is a problem for all of the perspectives discussed here, but it is a particularly acute one for a strongly situated view of knowledge. Lave and Wenger (1991) do not discuss much what a new entrant to a social practice brings with her, but it must include language, personal experience, the ability to learn, and usually the desire to participate in the activities of the social practice. All of these provide the foundation for learning but little in the way of transferable knowledge or skills.

Evans (1998) has argued from a post-structuralist situated cognition perspective that there are two things a person takes with them into a new social practice. First, their psychological make up – not in some unalterable essential sense – but with a formed if evolving psyche and a still open history of emotional sensitivities and responses. Second, a person also takes with them knowledge of signifiers and whole systems of signification, comprising language systems and tools and some of the associated skills. Thus a person in a new social context is in some sense the same person, with some corresponding emotional make-up and signifier resources, but open to the development of new facets of the self through new positionings and relationships. (see also Lerman 1999 for an elaboration of this perspective).

# REFERENCES

Ashlock, R. B. (1976) *Error patterns in Computation*, Columbus Ohio: Merrill.
Askew, M. and Wiliam, D. (1995) *Recent research in mathematics education 5-16*. London: Ofsted.
Assessment of Performance Unit (1979-1981) *Mathematical Development: Primary and Secondary Survey Reports*, London, HMSO.
Assessment of Performance Unit (1985) *A Review of Monitoring in Mathematics 1978 to 1982*, London: DES, HMSO.
Assessment of Performance Unit (1991) *APU Mathematics Monitoring (Phase 2)*, Slough: National Foundation for Educational Research (authors of report: D. Foxman, G. Ruddock, I. McCallum, I. Schagen).
Attisha, M. and Yazdani, M. (1984) An Expert System For Diagnosing Children's Multiplication Errors, *Instructional Science* 13(1): 79-92.
Ausubel, D. P. (1968) *Educational Psychology, a cognitive view*, New York: Holt, Rinehart and Winston.
Bacon, F. (1960) *The New Organon*, Indianapolis: Bobbs-Merrill.
Barclay, T. (1980) 'Buggy' *Mathematics Teaching*, 92, 10-12 (Sept. 1980).
Bateson, G. (1972) *Steps to an ecology of mind*. New York: Ballantine Books.
Bauersfeld, H. (1980) 'Hidden Dimensions in the So-Called Reality of a Mathematics Classroom', *Educational Studies in Mathematics* 11, 23-41.
Bauersfeld, H. (1995) 'The Structuring of the Structures', in Steffe and Gale (1995) 137-158.
Bell, A. W. (1989) Teaching for the test, *Times Educational Supplement* (27.10.89)
Bell, A. W., Küchemann, D. and Costello, J. (1983) *A Review of Research in Mathematical Education: Part A, Teaching and Learning*, NFER-Nelson, Windsor.
Berger, P. L. and Luckmann, T. (1966) *The Social Construction of Reality: A Treatise in the Sociology of Knowledge*, London: Penguin Books.
Bergeron, J. C., Herscovics, N. and Moser, J. M. (1986). Long Term Evolution of Students Conceptions: An Example from Addition and Subtraction. In Carss, M., Ed., (1986) *Proceedings of the Fifth International Conference on Mathematical Education*, Boston: Birkhauser: 275-277.
Biehler, R., Scholz, R. W., Straesser, R. and Winkelmann, B. Eds. (1994) *The Didactics of Mathematics as a Scientific Discipline*, Dordrecht: Kluwer.
Bishop, A. J. (1985) The Social Construction of Meaning: A Significant Development for Mathematics Education? *For the Learning of Mathematics*, Vol. 5, No. 1, 24-28.
Bishop, A., Ed. (1996) *International Handbook of Mathematics Education*, Dordrecht: Kluwer
Bloom, B. S. Ed. (1956) *Taxonomy of Educational Objectives 1, Cognitive Domain*, New York: David McKay.
Booth, L. (1982) 'Getting the answer wrong' *Mathematics in School* 11(2) 4- 6 (March 1982).
Booth, L. (1985) *Algebra: Children's Strategies and Errors*, Windsor: NFER- Nelson.
Bright, G. W. (1984) Computer Diagnosis of Errors, *School Science and Mathematics*, 84(3): 208-219.
Brown J. S., Collins, A. and Duguid, P. (1989) Situated Cognition and The Culture of Learning, *Educational Researcher*, 18(1): 32-42
Brown, G. and Desforges, C. (1979) *Piaget's Theory: A Psychological Critique*, London: Routledge and Kegan Paul.
Brown, J. and Van Lehn, K. (1982) Towards a Generative Theory of 'Bugs', in Carpenter, T. P. *et al.*, Eds., (1982) *Addition and Subtraction: A Cognitive Perspective*, Mahwah, New Jersey: Lawrence Erlbaum Associates.

Brown, J. S. and Burton, R. R. (1978) Diagnostic Models for Procedural Bugs in Basic Mathematical Skills, *Cognitive Science*, 2, 155-192.
Brown, M. and Küchemann, D. E. (1976, 1977, 1981) Is It An Add Miss? Parts 1, 2 and 3, *Mathematics In School* 5(5) 15-17 (1976), 8(1) 9-10 (1977), 10(1) 13-15 (1981).
Brownell, W. A. (1942) Problem Solving, *The Psychology of Learning* (41st NSSE Yearbook, Part II), Chicago: National Society for the Study of Education.
Bruner, J. (1960) *The Process of Education*, Cambridge, Massachusetts: Harvard University Press.
Bruner, J. (1964) *Towards a Theory of Instruction*, Cambridge, Massachusetts: Harvard University Press.
Bruner, J. S. (1974) *Beyond the Information Given: Collected Papers*. London: George Allen and Unwin.
Burton, L. (1984) *Thinking Things Through*, Oxford: Blackwell.
Carpenter, T. P., Moser, J. M. and Romberg, T. (1982) *Addition and Subtraction: A Cognitive Perspective*, Hillsdale, New Jersey: Lawrence Erlbaum.
Carraher, T. (1988) Street Mathematics and School Mathematics, in Borbas, A Ed. (1988) *PME - 12 Conference Proceedings*, Veszprem, Hungary, Vol. 1, 1-23.
Chaiklin, S. and Lave, J. (1993) *Understanding Practice: Perspectives on Activity and Context*. Cambridge: Cambridge University Press.
Champagne, A. D. (1992) Cognitive Research on Thinking in Academic Science and Mathematics: Implications for Practice and Policy, in Halpern (1992), 117-133.
Charles, R. I. (1985) The role of problem solving, *Arithmetic Teacher*, February 1985: 48-50.
Charles, R. I. and Lester, F. (1982) *Teaching Problem Solving*, New York: Wiley.
Cockcroft, W. H. Chair (1982) *Mathematics Counts*, London: Her Majesty's Stationery Office.
Cockcroft, W. H. Chair, (1982) *Mathematics Counts*, London: HMSO.
Davis, P. J. and Hersh, R. (1980) *The Mathematical Experience*, London: Penguin.
Davis, R. B., Maher, C.A. and Noddings, N., Eds. (1990) *Constructivist Views on the Teaching and Learning of Mathematics* (JRME Monograph 4), Reston, Virginia: National Council of Teachers of Mathematics
Denvir, B. and Brown, M. (1986a, b) 'Understanding of Number Concepts in Low Attaining 7-9 Year Olds: Parts I and II' *Educational Studies in Mathematics*, volume 17, 15-36 and 143-164.
Dept of Education and Science (1988) *Mathematics for ages 5 to 16*, London: DES/HMSO.
Descartes, R. (1931) *Philosophical Works*, Volume 1, Cambridge: Cambridge University Press (reprinted by Dover Press, New York, 1955).
Desforges, C. W. and Cockburn, A. (1987) *Understanding the Mathematics Teacher*, Basingstoke: Falmer Press.
Dessart, D. J. and Suydam, M. N. (1983) *Classroom Ideas From Research On Secondary School Mathematics: Algebra and Geometry*, Reston, Virginia: National Council of Teachers of Mathematics.
Dickson, L., Brown, M. and Gibson, D. (1984) *Children Learning Mathematics: A Teachers Guide To Recent Research*, Eastbourne: Holt Education.
Dienes, Z. P. (1960) *Building up Mathematics*, London: Hutchinson.
Dienes, Z. P. (1960) *Building Up Mathematics*, London: Hutchinson.
Dienes, Z. P. (1960a) *Number: Multibase Arithmetic Blocks and Powers Extension For MAB* (Mathematics Apparatus and Text), London: Educational Supply Association.
Dienes, Z. P. (1961) *Algebra: AEM* (Mathematics Apparatus and Text), London: Educational Supply Association.
Dienes, Z. P. (1964) *The Power of Mathematics*, London: Hutchinson.
Dienes, Z. P. (1966) *Mathematics in the Primary School*, London: Macmillan.
Dienes, Z. P. (1970) *The Six Stages in the Process of Learning Mathematics*, Windsor: NFER.

Dienes, Z. P. (1970) *The Six Stages in the Process of learning Mathematics*, Windsor: National Foundation for Educational Research.
Donaldson, M. (1978) *Children's Minds*, Glasgow: Fontana/Collins.
Driver, R. and Oldham, V. (1986) A Constructivist Approach to Curriculum Development, *Studies in Science Education* 13: 105-122.
Erlwanger, S. H. (1973) 'Benny's conception of Rules and Answers in IPI Mathematics', *Journal of Children's Mathematical Behaviour* 1(2), 8-25, 1973.
Ernest, P. (1983a) Mental Images, *Mathematics Teaching*, No. 103, 2-3.
Ernest, P. (1983b) Thinking of a Funny Line, *Mathematical Education for Teaching*, 4 (2) 30-40.
Ernest, P. (1984) Investigations, *Teaching Mathematics and its Applications*, 3(2),47-55.
Ernest, P. (1986) Mental Number Line Images', *Teaching Mathematics and its Applications* 5 (1) 1-5.
Ernest, P. (1988a) The mental number lines of a sample of adults', *International Journal of Mathematical Education in Science and Technology* 15 (3) 389-395.
Ernest, P. (1988b) What's the use of Logo?, *Mathematics in School*, 17 (1) 16-20.
Ernest, P. (1991a) *The Philosophy of Mathematics Education*, London: The Falmer Press.
Ernest, P. (1991b) Constructivism, The Psychology of Learning, and the Nature of Mathematics: Some Critical Issues, in *Proceedings of PME-15* (Italy), 2, 25-32. Reprinted in *Science and Education*, 2 (2), 1993, 87-93.
Ernest, P. (1994a) 'The Dialogical Nature of Mathematics', in Ernest, P. Ed. (1994) *Mathematics, Education and Philosophy: An International Perspective*, London, The Falmer Press, 33-48.
Ernest, P. (1994b) 'What is Social Constructivism in the Psychology of Mathematics Education?', Ernest, P. Ed. (1994) *Constructing Mathematical Knowledge: Epistemology and Mathematics Education*, London: The Falmer Press.
Ernest, P. (1996) *A Bibliography of Mathematics Education,* Exeter: University of Exeter School of Education. Available on WWW at http://www.ex.ac.uk/~PErnest/
Ernest, P. (1998) *Social Constructivism as a Philosophy of Mathematics*, Albany, NY: SUNY Press.
Ernest, P. (Forthcoming a) *Mathematics and Special Educational Needs*, Saarbrücken, Germany: Lambert Academic Publishing.
Ernest, P. (Forthcoming b) *The Mathematics Curriculum and Assessment*, Saarbrücken, Germany: Lambert Academic Publishing.
Ernest, P. (Forthcoming c) *Gender, Equal Opportunities and the Nature of Mathematics*, Saarbrücken, Germany: Lambert Academic Publishing.
Ernest, P. (Forthcoming d) *Research Methodology in Mathematics Education*, Saarbrücken, Germany: Lambert Academic Publishing.
Ernest, P. Ed. (1989) *Mathematics Teaching: The State of the Art*, London: Falmer Press.
Ernest, P. Ed. (1989) *Recent Developments in Primary Mathematics: A Course Reader*, University of Exeter, Exeter.
Evans, J. (1998) Boundary Crossing and Transfer, in Watson, A. Ed., *Situated Cognition and the Learning of Mathematics*, 1998, Oxford: Oxford University Department of educational Studies and York: QED Books (ISBN 1-85853-083-0).
Fernandez, M. L., Hadaway, N. and Wilson, J. W. (1993) Problem Solving: Managing It All, *Mathematics Teacher* 87(3) 195-199.
Fey, J. Y. (1996) Eclectic approaches to elementarization: Cases of curriculum construction in the United States, in Biehler, R. Scholz, R. W., Straesser, R. and Winkelmann, Eds. (1994) *The Didactics of Mathematics as a Scientific Discipline*, Dordrecht: Kluwer, 15-26.
Fischbein, E. (1994) The interaction between the formal, the algorithmic, and the intuitive components in a mathematical activity, in Biehler *et al.* (1994) 231-245.

Flavell, J. H. (1976) metacognitive aspects of problem solving, in Resnick, L. B. Ed. (1976) The Nature of intelligence, Hillsdale, New Jersey: Lawrence Erlbaum, 197-218.
Freire, P. (1972) *Pedagogy of the Oppressed*, London: Penguin Books.
Freudenthal, H. (1978) *Weeding and Sowing*, Dordrecht: Reidel.
Gagné, R. M. (1985) *The Conditions of Learning* (4th edition), New York: Holt, Rinehart and Winston
Galton, F. (1883) *Inquiries into Mental Faculties*, Reprinted in Everyman's Library, London: Dent and Sons.
Garofalo, J. and Lester, F. K. (1985) Metacognition, cognitive monitoring, and mathematical performance, *Journal for Research in Mathematics Education 16*(3): 163-176.
Ginsburg, H. (1977) *Children's Arithmetic: The Learning Process*, New York : Van Nostrand.
Ginsburg, H. P. Ed. (1983) *The Development of Mathematical Thinking*, New York: Academic Press.
Glasersfeld, E. von (1983) 'Learning as a Constructive Activity', in *Proceedings of the 5th PME-NA Conference*, Montreal: PME-NA, Vol. 1, 41-69.
Glasersfeld, E. von (1989) Constructivism in Education, in Husen, T. and Postlethwaite, N. Eds. (1989) *International Encyclopaedia of Education* (Supplementary Vol.), Oxford: Pergamon, 162-163.
Glasersfeld, E. von (1995) *Radical Constructivism: A Way of Knowing and Learning*. London, The Falmer Press.
Glasersfeld, E. von Ed. (1991) *Radical Constructivism in Mathematics Education*, Dordrecht: Kluwer.
Goldin, G. (1990) Epistemology, Constructivism, and Discovery Learning Mathematics, in Davis, R. B., Maher, C.A. and Noddings, N., Eds. (1990) *Constructivist Views on the Teaching and Learning of Mathematics* (JRME Monograph 4), Reston, Virginia: National Council of Teachers of Mathematics, 31-47.
Goldin, G. A. (1987) in Janvier, C. Ed. (1987) *Problems of Representation in the Teaching and Learning of Mathematics*, Hillsdale N J: Erlbaum.
Goldin, G. A. (1992) On developing a unified model for the psychology of mathematics education and problem solving, in Geeslin, W. and Graham, K. Eds (1992) *Psychology of Mathematics Education Conference 16 Proceedings*, Durham, New Hampshire, Volume 3, 235-261.
Goodchild, S. (1994). Blind Activity? *Proceedings of British Society for Research into Learning Mathematics Day Conference*, 22 Nov. 1993, Manchester.
Goodchild, S. (1997) *Unpublished PhD Thesis*, Exeter: University of Exeter School of Education.
Groner, R, Groner, M. and Bischof, W. F. (1983) *Methods of Heuristics*, Hillsdale, New Jersey: Erlbaum.
Grouws, D. A. Ed. (1992) *Handbook of Research on Mathematics Teaching and Learning*, New York: Macmillan.
Habermas, J. (1981) *The Theory of Communicative Action* (Trans. T. McCarthy), 2 volumes, Cambridge: Polity Press.
Halpern, D. F. (1992a) A Cognitive Approach to Improving Thinking Skills in the Sciences and Mathematics, in Halpern (1992) 1-14.
Halpern, D. F. Ed. (1992) *Enhancing Thinking Skills in the Sciences and Mathematics*, Hillsdale, New Jersey: Lawrence Erlbaum, 117-133.
Harré, R. (1979) *Social Being*, Oxford: Basil Blackwell.
Hart, K. (1978) 'Mistakes in Mathematics' *Mathematics Teaching*, 85, 38-40 (Dec. 1978).
Hart, K. (1985) *Ratio: Children's Strategies and Errors*, Windsor: NFER- Nelson.
Hart, K. (1989) There is Little Connection, in Ernest, P. Ed. (1989) *Mathematics Teaching: The State of the Art*, London: Falmer Press.

Hart, K. Ed. (1980) *Secondary Schoolchildren's Understanding of Mathematics*, London: Chelsea College.
Hart, K. Ed. (1981) *Children's Understanding of Mathematics: 11-16*, London: John Murray.
Hart, K. Ed. (1986) *The Chelsea Diagnostic Mathematics Tests* (Manual and 10 tests), Windsor: NFER-Nelson.
Hartzler, S. (1985) Mathematics DARES in Oklahoma City Public Schools, in Driscoll, M. and Confrey, J. Eds. *Teaching Mathematics*, Chelmsford, Massachusetts: Northeast Regional Exchange, 1985.
Hartzler, S. (1988) Details of and Results of Mathematics DARES Program in Oklahoma City Public Schools. *Unpublished papers.* Oklahoma City, Oklahoma: Curriculum Services Department, Oklahoma City Public Schools.
Haseman, K. (1991) The Use of Concept Maps to Explore Pupils' Learning Processes in Primary School Mathematics, in Furinghetti, F. Ed. *Proceedings of 15th International Conference on the Psychology of Mathematics Education*, Assisi, Italy, 2, 1991, 149-156.
Head, J. (1986) Research into 'Alternative Frameworks': promise and problems, *Research in Science and Technological Education, 4(2): 203-211.*
Her Majesty's Inspectorate (1985) *Mathematics from 5 to 16*, London: HMSO.
Hiebert, J. (1986a) Conceptual and Procedural Knowledge in Mathematics: an introductory analysis, in Hiebert (1986): 1-27.
Hiebert, J. Ed. (1986) *Conceptual and procedural knowledge: the case of mathematics*, Hillsdale, New Jersey: Erlbaum.
HMI (1987) *Mathematics from 5 to 16: the Responses to Curriculum Matters 3*, HMSO, London
Holt, M. and Dienes, Z. P. (1973) *Lets Play Maths*, London: Penguin.
Hoyles, C. and Noss, R. Eds. (1992) *Learning mathematics and Logo*, Cambridge, Massachusetts: Massachusetts Institute of Technology Press.
Hoyles, C. and Sutherland, R. (1989) *Logo Mathematics in the Classroom*, Blackwell, Oxford.
Huerta, M. P. (1996) using concept mapping to analyse students' relationships between quadrilaterals, paper presented at *20th International Psychology of Mathematics Education Conference*, University of Pernambuco, Recife, Brazil, July 1996.
Hughes, M. (1986) *Children and Number*, Basil Blackwell, Oxford.
Jensen, R. J. Ed. (1993) *Research Ideas for the Classroom: Early Childhood Mathematics*, New York: Macmillan.
Johnson, D. C. Ed. (1989) *Children's Mathematical Frameworks*, Windsor: NFER-Nelson.
Kent, D. (1978) Some processes through which mathematics is lost, *Educational Research*, 21(1): 27-35.
Kent, D. (1979) More about the processes through which mathematics is lost, *Educational Research*, 22(1): 21-31.
Kerslake, D. L. (1985) *Fractions: Children's Strategies and Errors*, Windsor: NFER- Nelson.
Kieren, T. E. and Pirie, S. E. B. (1991) Recursion and the Mathematical Experience, in Steffe (1991), 78-101.
Kilpatrick, J. (1992) A History of Research in Mathematics Education, in Grouws (1992) 3-38.
Kitcher, P. (1984) *The Nature of Mathematical Knowledge*, New York: Oxford U. Press.
Klingele, W. E. and Reed, B. W. (1984) An examination of an incremental approach to mathematics, *Phi Delta Kappan*, 1984 (June): 712-713.
Kuhn, T. S. (1970) *The Structure of Scientific Revolutions*, Chicago: Chicago University Press
Laing, R. A. (1970) *Relative affects of massed and distributed scheduling of topics on homework assignments of eighth grade mathematics students*, Unpublished PhD dissertation, Kalamazoo, Michigan: Western Michigan University.
Lakatos, I. (1976) *Proofs and Refutations*, Cambridge: Cambridge University Press.

Lakatos, I. (1978) *The Methodology of Scientific Research Programmes* (Philosophical Papers Volume 1), Cambridge: Cambridge University Press.

Lange, J. de (1996) Using and Applying Mathematics in Education, in Bishop, A., Ed., (1996) *International Handbook of Mathematics Education*, Dordrecht: Kluwer, 49-97.

Lankford, F. (1972) *Some computational strategies of seventh grade pupils*, Virginia: University of Virginia, National Centre for Educational Research.

Larcombe, T. (1985) *Mathematical Difficulties in the Secondary School*, Open University Press, Milton Keynes

Lave, J. (1988) *Cognition in Practice*, Cambridge: Cambridge University Press.

Lave, J. and Wenger, E. (1991) *Situated Learning: Legitimate Peripheral Participation*, Cambridge: Cambridge University Press.

LeBlanc, J. F. (1982) A Model for Elementary Teacher Training in Problem Solving, in Lester and Garofalo (1982): 111-115.

Leinhardt, G. (1987) The Development of an Expert Explanation: An Analysis of a Sequence of Subtraction Lessons, *Cognition and Instruction*, 4: 225-282.

Leinhardt, G. and Smith, D. A. (1985) Expertise in Mathematics Instruction: Subject Matter Knowledge, *Journal of Educational Psychology*, 77(3), 247-271.

Lerman, S. (1999) Culturally Situated Knowledge and the Problem of Transfer in the Learning of Mathematics, in Burton, L. Ed. (1999) *Learning Mathematics, from Hierarchies to Networks*, London, The Falmer Press.

Lester, F. K. and Garofalo, J. Eds. (1982) *Mathematical Problem Solving*, Philadelphia, Pennsylvania: Franklin Institute Press.

Lester, Jr., F. K., Masingila, J. O., Mau, S. T., Lamdin, D. V., Santos, V. M. P. dos and Raymond, A. M. (1994) Learning how to teach via problem solving, in D. Aichele, Ed., (1994) *Professional development for teachers of mathematics* (Yearbook of the National Council of Teachers of Mathematics), Reston, Virginia: National Council of Teachers of Mathematics, 152-166.

Lindquist, M. M. (1988) *Selected Issues In Mathematics Education*, Berkeley, California: McCutchan.

Luria, A. R. (1977) *The Mind of a Mnemonist*, London: Penguin Books.

Malone, J. and Dekkers, J. (1984) The concept map as an aid to instruction in science and mathematics, *School Science and Mathematics* 84(3) 220-231.

Mason, J. with Burton, L. and Stacey, K. (1982) *Thinking Mathematically*, London: Addison Wesley.

Mayer, R. E. (1986) Mathematics, in Dillon, R. F. and Sternberg, R. J. Eds. (1986) *Cognition and Instruction*, New York: Academic Press, 127-154.

Mellin-Olsen, S. (1981) Instrumentalism as an Educational Concept, *Educational Studies in Mathematics*, Vol. 12, 351-367

Miller, G. (1956) 'The magic number seven, plus or minus two: Some limits on our capacity for processing information', *Psychological Review*, 63: 81-97.

National Curriculum Council (1991) *Mathematics in the National Curriculum*, London: DES.

National Curriculum Mathematics Working Group (1987) *Interim Report*, London: DES.

Niss, M. (1996) Goals of Mathematics Teaching, in Bishop, A., Ed. (1996) *International Handbook of Mathematics Education*, Dordrecht: Kluwer, 11-47.

Noss, R. (1983) Doing Maths while leaning Logo, *Mathematics Teaching*, No. 104 (September 1983).

Noss, R. (1988) The Computer as a Cultural Influence in Mathematical Learning, *Educational Studies in Mathematics* 19(2), 251-268.

Novak, J. D. (1990) Concept mapping: A useful tool for science education, *Journal of Research in Science Teaching*, 27: 937-952.

Novak, J. D. Ed.(1987) *Proceedings of the Second International Seminar on Misconceptions and Educational Strategies in Science and Mathematics,* July 1987 (3 Volumes), Ithaca, New York: Cornell University

Novak, J. D. et al. (1983) The use of concept mapping and knowledge Vee with junior high school science students, *Science,* 67(5): 625-645.

Orton, A. (1992) *Learning Mathematics* (2nd edition), London: Cassell.

Owens, D. T. Ed. (1993) *Research Ideas for the Classroom: Middle Grades Mathematics,* New York: Macmillan.

Papert, S. (1980) *Mindstorms: Children Computers and Powerful Ideas,* Brighton: Harvester.

Perkins, D. N. and Salomon, G. (1989) Are cognitive skills context bound? *Educational Researcher* 18(1): 16-25.

Pfundt, H. and Duit, R. (1988) *Bibliography: Students' Alternative Frameworks and Science Education* (IPN Reports-in-Brief 34) Kiel, Federal Republic of Germany: Institute for Science Education, University of Kiel.

Piaget, J. (1952) *The Child's Conception of Number,* New York: Norton.

Piaget, J. (1953) How Children Form Mathematical Concepts, *Scientific American,* November 1953

Piaget, J. (1969) *The Psychology of The Child,* New York: Basic Books.

Piaget, J. (1971) *The Construction of Reality in the Child* (Translated by M. Cook), New York: Basic Books.

Piaget, J. (1972) *Psychology and Epistemology: Towards a Theory of Knowledge,* Harmondsworth: Penguin

Pirie, S. (1987) *Mathematical Investigations in Your Classroom,* Basingstoke: Macmillan.

Plunkett, S. (1979) Decomposition and all that Rot, *Mathematics in School,* 8, 2-7.

Polanyi, M. (1958) *Personal Knowledge,* London: Routledge & Kegan Paul.

Polya, G. (1945) *How to Solve it,* Princeton, New Jersey Princeton University Press.

Polya, G. (1954) *Mathematics and Plausible Reasoning (Volume 1: Induction and Analogy in Mathematics, Volume 2: Patterns of Plausible Inference),* Princeton, N J: Princeton University Press.

Polya, G. (1981) *Mathematical Discovery: On Understanding Learning and Teaching Problem Solving* London: Oxford: Wiley

Popper, K. (1959) *The Logic of Scientific Discovery,* Hutchinson, London.

Popper, K. R. (1979) *Objective Knowledge* (Revised Edition), Oxford: Oxford University Press

Powell, A. B. and Frankenstein, M. (1997) *Ethnomathematics: Challenging Eurocentrism in Mathematics,* Albany, New York: SUNY Press.

Rees, R. and Barr, G. (1984) *Diagnosis and Prescription - Some Common Maths Problems,* London: Harper.

Resnick, L. and Ford, W. W. (1981) *The Psychology of Mathematics for Instruction,* London: Erlbaum.

Resnick, L. B. (1983) A Developmental Theory of Number Understanding, in Ginsburg (1983) 110-151.

Resnick, L. B. and Ford, W. W. (1981) *The Psychology of Mathematics for Instruction,* Erlbaum, London.

Reynolds, J. H. and Glaser, R. (1964) 'Effects of repetition and spaced review upon retention of a complex learning task' *Journal of Educational Psychology,* 55: 297-308.

Riding, R. J. (1977) *School Learning: Mechanisms and Processes,* London: Open Books.

Robitaille, D. F. and Garden, R. A. Eds (1989) *The IEA Study of Mathematics II: Contexts and Outcomes of School Mathematics,* Oxford: Pergamon.

Robitaille, D. F. and Travers, K. J. (1992) International Studies of Achievement in mathematics, in Grouws, D. A. Ed. (1992) *Handbook of Research on Mathematics Teaching and Learning*, New York: Macmillan, 687-709.

Rogoff, B. and Lave, J. Eds. (1984) *Everyday Cognition: Its development in a social context*, Cambridge, Massachusetts: Harvard University Press.

Rorty, R. (1979) *Philosophy and the Mirror of Nature*, Princeton, New Jersey: Princeton University Press.

Ryle, G. (1949) *The Concept of Mind*, London: Hutchinson.

Saxe, G. B. (1991) *Culture and Cognitive Development: Studies in Mathematical Understanding*, Hillsdale, New Jersey: Erlbaum.

Schoenfeld, A. (1985) *Mathematical Problem Solving*, New York: Academic Press.

Schoenfeld, A. (1992) Learning to Think Mathematically, in Grouws (1992) 334-370.

Schoenfeld, A. Ed. (1987) *Cognitive Sciences and Mathematics*, Hillsdale, New Jersey: Erlbaum.

Schunert, J. (1951) 'The association of mathematics achievement with certain factors resident in the teacher, in the pupil, and in the school', *Journal of experimental Education* 19.

Seaborne, P. L. (1975) *An Introduction to the Dienes Mathematics Program*, London: London University Press.

Sharron, H. (1987) *Changing children's minds: Feuerstein's revolution in the teaching of intelligence*, Bristol: Souvenir Press.

Skemp, R. R. (1971) *The Psychology of Learning Mathematics*, Penguin, Harmondsworth.

Skemp, R. R. (1976) 'Relational Understanding and Instrumental Understanding' *Mathematics Teaching*, No. 77, 20-26.

Slesnick, T. (1982) Algorithmic Skill vs. Conceptual Understanding, *Educational Studies in Mathematics* 13: 143-154.

Smith, D. M. (1987) An evaluation of Saxon's algebra text, *Journal of Experimental Education*, 81(2): 97-102.

Steffe, L. P. and Gale, J. Eds (1995) *Constructivism in Education*, Hillsdale, New Jersey: Lawrence Erlbaum Associates.

Steffe, L. P. Ed. (1991) *Epistemological Foundations of mathematical Experience*, New York: Springer-Verlag.

Steffe, L. P., Nesher, P., Cobb, P., Goldin, G. A, and Greer, B. Eds. (1996) *Theories of Mathematical Learning*, Mahwah, New Jersey: Erlbaum,

Thomas, W. E. and Grouws, D. A. (1984) Inducing cognitive growth in concrete-operational college students, *School Science and Mathematics* 84: 223-243.

Thyne, J. M. (1954) *Patterns of Error*, London: University of London Press.

Underhill, R. G., Uprichard, A. E. and Heddens, J. W. (1980) *Diagnosing Mathematical Difficulties*, Columbus, Ohio: Merrill.

Van Lehn, K. (1983) On the representation of procedures in repair theory, in Ginsburg (1983), 201-252.

Vygotsky, L. S. (1979) The Instrumental Method in Psychology, in Wertsch, J. V. (1979) *The Concept of Activity in Soviet Psychology*, Armonk, New York: M. E. Sharpe, Inc., 134-143.

Vygotsky, L. S. (1978) *Mind in Society; The development of the higher psychological processes*, Cambridge, Massachusetts: Harvard University Press.

Vygotsky, L. S. (1986) *Thought and Language*, Translated by A. Kozulin, Cambridge, Massachusetts: Massachusetts Institute of Technology Press. (First English translation 1962).

Wagner, S. and Kieran, C. Eds. (1989) *Research Issues in the Learning and Teaching of Algebra*, Reston, Virginia: National Council of Teachers of Mathematics and Lawrence Erlbaum Associates.

Walkerdine, V. (1984) Developmental Psychology and the Child-Centred Pedagogy: The Insertion of Piaget into Early Education, In Henriques, J., Holloway, W., Urwin, C., Venn, C. and Walkerdine, V. (1984) *Changing The Subject*, London: Methuen, 153-202.
Ward, M. (1979) *Mathematics and the 10-year-old*, Evans/Methuen, London.
Wertsch, J. V. (1991) *Voices of the Mind*, London: Harvester Wheatsheaf.
Whetton, C. and Hagues, N. (1985) 'Some Modern Approaches to Testing' *Assessment and Evaluation: Proceeding of the 1985 NFER Members' Conference*, Windsor: NFER.
Wilson, P. A., Ed. (1993) *Research Ideas for the Classroom: High-School Mathematics*, New York: Macmillan.
Wittgenstein, L. (1953) *Philosophical Investigations* (trans. G. E. M. Anscombe), Oxford: Basil Blackwell.
Wittman E, (1985) Winter's theory of practice, *Proceedings of 37th CIEAM Conference*, 1985.
Wolff, P. (1963) *Breakthroughs in Mathematics*, New York: New American Library.
Wood, T., Cobb, P. and Yackel, E. (1995) 'Reflections on Learning and Teaching Mathematics in Elementary School', in Steffe and Gale (1995) 401-422.

## VDM publishing house ltd.

# Scientific Publishing House
### offers
# free of charge publication

of current academic research papers, Bachelor´s Theses, Master's Theses, Dissertations or Scientific Monographs

If you have written a thesis which satisfies high content as well as formal demands, and you are interested in a remunerated publication of your work, please send an e-mail with some initial information about yourself and your work to *info@vdm-publishing-house.com*.

**Our editorial office will get in touch with you shortly.**

**VDM Publishing House Ltd.**
Meldrum Court 17.
Beau Bassin
Mauritius
www.vdm-publishing-house.com

Printed in Great Britain
by Amazon.co.uk, Ltd.,
Marston Gate.